Geometry at the Quantum Scale

Marcos D. Maia • Edmundo M. Monte

Geometry at the Quantum Scale

Understanding the Geometric Language of Relativistic Quantum Mechanics

Springer

Marcos D. Maia
Institute of Physics
Universidade Brasilia
Brasília, Brasília, Brazil

Edmundo M. Monte
Department of Physics
Paraiba University
João Pessoa, Paraíba, Brazil

ISBN 978-3-031-61196-4 ISBN 978-3-031-61197-1 (eBook)
https://doi.org/10.1007/978-3-031-61197-1

Mathematics Subject Classification: 51P05, 81V17, 81T13, 14R99

This Springer imprint is published by the registered company Springer Nature Switzerland AG
The registered company address is: Gewerbestrasse 11, 6330 Cham, Switzerland

Preface

This book is about Quantum Cosmology, a discipline describing the universe as a result of all known fundamental interactions, including classical and quantum gravity and the three known gauge forces, which compose the Standard Model of Fundamental Interactions (The Standard Model for short). Therefore, in contrast with the present Classical Cosmology associated with only classical gravitation, Quantum Cosmology involves all known fundamental interactions defined at the quantum scale, including quantum gravity. The result is an interesting 11-dimensional quantum geometry defined by the Standard Model of Interactions.

The Standard Model of Interactions owes its strength to the fact that instead of being a postulated model theory of everything, it is rather based on the collected data that started begin collected from the beginning of the twentieth century with the electromagnetic interaction, to the high energy physics established in the present, leading us to a solid observational basis from which we may construct a realistic model of whatever composes the universe where we live. The gravitational force must be present in the Standard Model of Interactions represented by a massless spin-2 particle. Yet, in spite of the intense theoretical efforts made in that direction during the past century, we still do not have a unique model of quantum gravity in the sense defined by Einstein. Rather, today we count dozens of such models which is even worse. The main difficulty to quantize the gravitational field resides in its observational property: The observers (actually we) are of classical nature and as such cannot just walk into the quantum realm, much less carrying a general covariant local metric stick.

The end purpose of this book is to present a geometrical formulation of the Standard Model, including Einstein's Relativistic gravitation, such that it is consistent with the renormalization of the gravitational field at the same time that it will also eliminate the hierarchy of the gravitational field of General Relativity, with respect to the gauge theories. The renormalization is obtained from an infinite sequence of perturbations of the Minkowski metric by powers of a massless spin-2 quantum field, leading to a general covariant quantum gravitational field in the same geometrical sense defined by Einstein, with the difference that it is now defined at the quantum scale of observations.

Consequently, and similarly to what happens in gauge field theory, the observations of quantum gravity must use an appropriate affine connection, associated with the group of diffeomorphisms of the space-time, in conformity with Einstein's principle of general covariance. The elimination of the hierarchy of the fundamental interaction requires that the classic Newtonian gravitational coupling constant of the gravitational field must be adjusted to the quantum scale of observations. Such change implies also that the classic symmetry mixing problem of particle physics must be revised to maintain the Poincaré and gauge symmetries together at the quantum scale. We show how the solution of that classic problem can be obtained with the Higgs mechanism.

The resulting geometry at the quantum scale involving quantum gravitation and the three gauge interactions produce an interesting 11-dimensional general covariant Riemannian geometry defined at the quantum scale, whose curvature tensor provides the field equations describing the Standard Model of Interactions, expressed in terms of the three fundamental forms of geometry: the first fundamental form (the metric) describing the gravitational potential of the quantum gravitational field, the third fundamental form describing the three known gauge potentials and quite surprisingly, the second fundamental form describing a new quantum topological phenomenon—a relativistic version of the Ricci flow and a relativistic version of the Poincaré conjecture. Also, we will see that the 11-dimensional Einstein-Hilbert Lagrangian decomposes into the four-dimensional Einstein-Hilbert Lagrangian for Einstein's gravitational field, plus the four-dimensional Yang-Mills Lagrangian. The resulting 11-dimensional geometry appears to be an extension of Donaldson's four-manifold theory defined at the quantum domain.

Brasília, Brazil Marcos D. Maia
João Pessoa, Brazil Edmundo M. Monte

Acknowledgments

We would like to express our warmest thanks to Edmundo Matheus C. Monte, Bianca C. Monte, and Alison Silva for their nice artwork helping to make our quantum geometry more appealing to our senses. Our gratitude to all colleagues, teachers, collaborators, students, actually everyone who directly or indirectly contributed to our understanding about the nature of things. We also would like to express our thanks to the Springer editorial staff for the suggested improvements on the composition of the somewhat complex nature of our book. Finally, we thank the permissions given by the authors of some excellent drawings displayed through the internet. We especially thank the Bodleian library at Oxford for permission to use the drawing by Rustischello de Pisa, showing the Polo's departing from Venice, providing an excellent example of how the quantum space-time could look from a distance, without the use of a general covariant affine connection.

Contents

List of Figures

Chapter 1
Gauge Forces

Abstract We present a short overview of the development of gauge forces, from Maxwell's electromagnetic theory to the presently known Yang-Mills theory. Although this is well known, we detail here the proof of Noether's theorem on the conservation laws for global and local gauge symmetries, often neglected in regular disciplines. We will show how Emmy Noether cleverly introduced new terms, which were later identified with the affine connections of the quantum geometry of the gauge fields.

The three fundamental forces of nature known as gauge forces represent a generalization of the experimental and mathematical studies on electricity and magnetism initiated by Charles A. Coulomb, Michael Faraday, Jean-Baptiste Biot, Félix Savart, and André-Marie Ampère, culminating just around the turn of the twentieth century in the electromagnetic theory synthesized by James Clerk Maxwell[1] [1].

The original formulations of the electromagnetic equations were written in a three-dimensional Euclidean space, together with the absolute time, defined by Newton in the seventeenth century, resulting in what is called *the Galilean space-time*, the simplest example of a "space-time," using Newton's definition of absolute time: If t is an absolute time, then $t' = at + b$, where $a \neq 0$ and b are real numbers, is another absolute time. The expressions for the electric field **E** and the magnetic field **B** are

$$\mathbf{E} = -\nabla\phi - \frac{1}{c}\frac{\partial \mathbf{A}}{\partial t}, \qquad \mathbf{B} = \nabla \wedge \mathbf{A}, \tag{1.1}$$

[1] It should be noted that the vector calculus that we use today had not been yet developed at Maxwell's time. Instead, Maxwell used Hamilton's noncommutative quaternion algebra. Here we shall use the present vector calculus formulation, where the expression $< \mathbf{a}, \mathbf{b} >$ denote the scalar product in the three-dimensional Euclidean space and $\nabla\phi = \sum_{i=1}^{3} \frac{\partial \phi}{\partial x^i}$ denotes the gradient of ϕ. The speed of light c is a constant. We will see later on that at the quantum scale the use of quaternions is more consistent.

M. D. Maia, E. M. Monte, *Geometry at the Quantum Scale*,
https://doi.org/10.1007/978-3-031-61197-1_1

where the electric field is derived from the electric field potential ϕ and the magnetic field is derived from the magnetic field potential $\mathbf{A}$. From these definitions, it follows that these potentials are solutions of two homogeneous equations, known, respectively, as the Faraday and the Biot-Savart equations

$$\nabla \wedge \mathbf{E} + \frac{1}{c}\frac{\partial \mathbf{B}}{\partial t} = 0, \qquad <\nabla, \mathbf{B}>= 0. \tag{1.2}$$

In addition, these potentials satisfy two nonhomogeneous equations

$$\nabla^2\phi - \frac{1}{c^2}\frac{\partial^2\phi}{\partial t^2} = -4\pi\rho, \quad \nabla^2\mathbf{A} - \frac{1}{c^2}\frac{\partial^2\mathbf{A}}{\partial t^2} = -\frac{4\pi}{c}\mathbf{j}, \tag{1.3}$$

known as the Coulomb and Ampère equations, respectively. However, it was only by the turn of the century that those equations were consistently put together by Maxwell, in what turned out to be the first successful unification of the then known electric and magnetic forces to the electromagnetic field, described by the equations

$$\begin{cases} <\nabla, \mathbf{B}>= 0, \\ \nabla \wedge \mathbf{E} + \frac{1}{c}\frac{\partial \mathbf{B}}{\partial \tau} = 0, \\ <\nabla, \mathbf{E}>= 4\pi\rho, \\ \nabla \wedge \mathbf{B} - \frac{1}{c}\frac{\partial \mathbf{E}}{\partial \tau} = \frac{4\pi}{c}\mathbf{J}, \end{cases} \tag{1.4}$$

where ρ represents the electric charge density and $\mathbf{J}$ represents the electric current vector. The last two equations are derived from the electromagnetic Lagrangian (using $c = 1$).

$$\mathscr{L}_{\text{EM}} = \frac{<\mathbf{E}, \mathbf{E}> - <\mathbf{B}, \mathbf{B}>}{8\pi} - \rho\phi - <\mathbf{J}, \mathbf{A}> . \tag{1.5}$$

Accordingly, the above equations are considered as dynamical equations. On the other hand, the first two equations (often referred to as the homogenous equations) can be seen to be just algebraic identities.

The expressions in (1.4) were written in the Galilean space-time which is a four-dimensional manifold.[2] A relevant fact of the concept of manifold is that they can be mathematically described by a set of parameters (or coordinates), which are related to one another by specific transformations, keeping the Eqs. (1.4) written

[2] The name manifold was used as a translation by W.K Clifford of the original concept "Maningfaltigkeit" introduced by Emmanuel Kant [2]. The translation to "variety" may sound more suitable, in the sense that a manifold may also mean the same thing folded in different shapes, while variety may include different things folded in different shapes.

in the same form, composing a *group of symmetries*. Thus, for example, in the case of Maxwell's equations, all observers associated with coordinates (x) and a time parameter (t) define a space and time location (x, t), such that they will interpret Maxwell's equations in the same way. This is in essence the base of Einstein's relativistic concept introduced in 1905.

Before the relativistic era, the absolute time parameter defined by Newton was separated from the space coordinates, such that Maxwell's equations were supposed to look the same as when they were transformed as

$$\begin{cases} x^i = a^i_j x^j + c^i t + d^i, & i, j = 1, 2, 3 \\ t' = at + b, \end{cases} \tag{1.6}$$

where the coefficients a^i_j, c^i, d^i, a, and b are such that the above transformations do not alter the equations in (1.4). The composition of such transformations forms a group, called the Galilean group of symmetry of the Galilean space-time.

Maxwell noted that the above equations led to wave equations for the electromagnetic field written as **E** and **B** as follows [3, 4]:

$$\Box^2 \mathbf{E} = \left(\nabla^2 - \frac{1}{c^2}\frac{\partial^2}{\partial t^2}\right)\mathbf{E} = -4\pi\left(\nabla\rho + \frac{\partial \mathbf{J}}{\partial t}\right), \tag{1.7}$$

$$\Box^2 \mathbf{B} = -\left(\nabla^2 - \frac{1}{c^2}\frac{\partial^2}{\partial t^2}\right)\mathbf{B} = -4\pi\nabla \wedge \mathbf{J}. \tag{1.8}$$

These electromagnetic waves propagate along a direction orthogonal to the planes defined by the vectors **E** and **B**, characterized as polarized plane waves.

However, by the end of the nineteenth century all known wave motions were manifested through the vibrations of some elastic material like air, water, metals, etc. Therefore, it was generally thought that the electromagnetic waves also required the existence of a *transport media*, an elusive substance that was called the "physical aether." This prompted the development of an experiment to verify the existence and properties of such aether.

The verification was accomplished through an experiment designed by Albert A. Michelson and Edward W. Morley using an adaptation of an earlier experiment to measure the speed of light in water performed by Hypollite Fizeau in 1676. Instead of water Michelson and Morley considered the flow of the "aether" on the surface of the Earth, as it rotates around the Sun (Fig. 1.1).

Michelson and Morley reasoned that the Earth's motion immersed in such aether at the speed of 30 kilometers per second (the translational velocity to the Earth rotation around the Sun) in principle would affect the propagation of light. The differences between the speed of light in the direction of the Earth motion and the propagation of light in the orthogonal direction would show the expected evidence

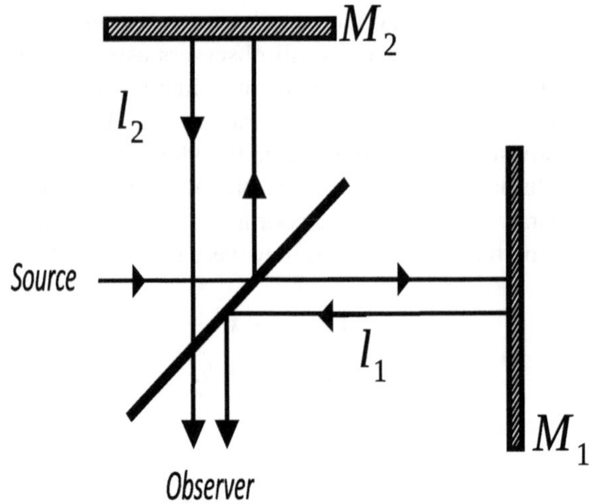

Fig. 1.1 The Michelson-Morley interferometer

for the existence of the aether. The experiment could also be checked with the opposite direction of the aether flow 6 months later.

The measurement was made through an interferometer composed of two pieces of glass separated by a very thin transparent media (a basic etalon), showing a diffraction pattern: After many different attempts to obtain a nonzero measure under different conditions (including high mountains and balloons at high altitudes), the experiment did not produce any appreciable difference in the diffraction pattern, *leading to the conclusion that the supposed aether did not exist and that light propagates with constant velocity c, assumed to be one of the fundamental constants of nature.* In other words, the negative result obtained by the Michelson-Morley experiments suggested that the electromagnetic waves were an intrinsic fundamental characteristic of the electromagnetic field, telling that the propagation of light does not depend on a mechanical support. That means that light propagates through the vacuum, as was shown later directly by H. Hertz in 1886.

Actually, we may say that the aether was replaced by a mathematical concept proposed by H. A. Lorentz when he found that Maxwell's equations were invariant under a six-parameter group of transformations of coordinates which is more general than the Galilean group (1.6), known today as the Lorentz group. That group was latter on extended by H. Poincaré to include four translation parameters, leading to a more general group, usually referred to as the "Poincaré group" of symmetries of the electromagnetic field.

The definition of a metric geometry, whose metric would be invariant under the Poincaré group, was defined only in 1908 by H. Minkowski, which gave its present name, the space-time manifold, to the four-dimensional Minkowski space-time of Einstein's Special Relativity Theory in 1905 [5].

As we see, the electromagnetic theory was the main motivation for the definition of today's notion of relativistic space-time, in place of the Newtonian concept of absolute space-time.[3] Therefore, using the convention that the speed of light is 1 ($c = 1$), the components of the Minkowski metric in four-dimensional orthogonal coordinates will be written as

$$(\eta_{\mu\nu}) = \begin{pmatrix} 1 & 0 & 0 & 0 \\ 0 & 1 & 0 & 0 \\ 0 & 0 & 1 & 0 \\ 0 & 0 & 0 & -1 \end{pmatrix}. \tag{1.9}$$

A review of the evolution of the concept of space-time can be found in Penrose's Adams Prize essay [6].

In order to understand further implications of Einstein's Special Relativity, we translate the original formulation of Maxwell's equations given by (1.4) originally written in the Galilean space-time to the Minkowski four-dimensional space-time as follows: Define a basis of the Minkowski space-time in which the electric potential ϕ and the magnetic vector potential $\mathbf{A}$ are seen as components of a four-dimensional potential, expressed by its components as $\mathbf{A} = (A_1, A_2, A_3, -\phi)$, and defining the Maxwell tensor $\mathbf{F}$ with contra-variant components $F_{\mu\nu} = (\partial_\mu A_\nu - \partial_\nu A_\mu)$ given by the 4×4 matrix

$$\mathbf{F} = \begin{pmatrix} 0 & B_3 & -B_2 & E_1 \\ -B_3 & 0 & B_1 & E_2 \\ B_2 & -B_1 & 0 & E_3 \\ -E_1 & -E_2 & -E_3 & 0 \end{pmatrix},$$

and whose covariant components are given by $F^{\mu\nu} = \eta^{\mu\rho}\eta^{\nu\sigma}F_{\rho\sigma}$.

Using the above tensor notation the electromagnetic Lagrangian simplifies to

$$\mathscr{L}_{\text{Maxwell}} = \frac{1}{4}F^{\mu\nu}F_{\mu\nu} - \frac{1}{c}J^\mu A_\mu. \tag{1.10}$$

In the tensor notation the four Maxwell equations become simply

$$F^{\mu\nu}{}_{,\nu} = 4\pi J_\mu A^\mu, \quad F^{*\mu\nu}{}_{,\nu} = 0, \tag{1.11}$$

[3] We will use the following convention: Unless specifically said in contrary, Greek indices like $\mu,\ \nu,\ \rho,\ \sigma, \ldots$ run from 1 to 4. Lowercase Latin indices like $a,\ b,\ c \ldots i,\ j,\ k \ldots$ generally run from 1 to 3, referring to the three-dimensional space. The four-dimensional space-time signature is taken to be $(+, +, +-)$, so that the relativistic time corresponds to the coordinate x_4. In a higher dimensional manifold, capital Latin indices like $A,\ B,\ C, \ldots$ run from 1 to N, regardless of its signature. Einstein's summation convention is used throughout, meaning that unless specified otherwise, all repeated upper and lower indices always add up. A matrix can be seen as a particular mixed tensor, with one contravariant index and one covariant index.

where we have denoted $F^{*\mu\nu} = \epsilon^{\mu\nu\rho\sigma} F_{\rho\sigma}$, called the *dual* of $F_{\rho\sigma}$, and $J_\mu = (J_1, J_2, J_3, \rho)$ is the 4-current vector.[4]

1.1 Gauge Symmetry

Gauge field theory emerged in 1918, when Hermann Weyl was studying the invariance of Maxwell's equations under some transformation of the potentials given by

$$\begin{cases} \mathbf{A}' = \mathbf{A} + \nabla\theta, \\ \phi' = \phi - \frac{1}{c}\frac{\partial\theta}{\partial t}. \end{cases} \tag{1.12}$$

Weyl noted that the above transformations do not follow from a coordinate transformation. In fact, they suggest the existence of a new kind of symmetry of Maxwell's equations, not related to the invariance of the metric of the Minkowski space-time. Indeed, Weyl found that those transformations do not follow from a Lorentz transformations and that they depend on a new independent parameter θ which may also be a function of the space-time coordinates. As such, these new transformations could in principle have new physical implications.

Rather, these new transformations (1.12) are related to a measure of a force derived from the potential. Weyl called these transformations gauge transformations, because they act like as if "setting a new standard of measurement. In the cases where the θ are functions of the space-time coordinates, they are called local gauge transformations, and otherwise they are called as global gauge transformations.

The gauge transformations of the electromagnetic field define a one-parameter Lie group called the *electromagnetic gauge group*. It is like the group of rotations $SO(2)$ acting on some two-dimensional plane space, not to be confused with a plane of the space-time. Since this group is isomorphic[5] to the unitary complex group $U(1)$, then Weyl suggested that these gauge transformations could be associated with the quantum theory of the electromagnetism (quantum electrodynamics). Thus, the isomorphism between the gauge group $U(1)$ and the group of rotations $SO(2)$ is not to be confounded with a group of rotations in the space-time. Rather it is a rotation in some other geometric space within the quantum domain. Thus, in order to avoid confusion, we shall refer to such space as the *internal quantum space of the* $U(1)$. Note that at the time when Weyl discovered the gauge symmetry, quantum theory was still in its infancy, and the notion of a space defined at quantum scale was not obvious.

---

[4] The expression $\epsilon^{\mu\nu\rho\sigma}$ is the four-dimensional complete antisymmetrization symbol: It is zero for any two equal indices and changes sign for any odd change of the order 1234.

[5] A general theorem tells that all Lie groups with the same number of parameters are isomorphic between themselves.

In order to get further insight on the physical significance of the new symmetry, Weyl asked his assistant, Emmy Noether, to find out what would be the invariant quantities associated with those gauge transformations. In 1919 Noether delivered a far more general result, when she derived the invariant quantities associated with a Lie group of symmetries of a given action integral defined by a Lagrangian functional.

The mathematical definition of a *group of transformations* is similar to that of a manifold, where each element of the group, say $a, b, c\ldots$, is defined by a certain number of parameters, such as $a = f(\theta_1, \theta_2, \theta_3 ...)$, $b = (\theta'_1, \theta'_2, \theta'_3 \ldots)$, etc. The difference is that as a group the composition of any two elements, say $c = ab$, necessarily involves a relation between the respective parameters, which can be generically described as $\theta''' = \phi(\theta', \theta'')$. In a differentiable manifold such a relation is given by a differentiable function. However, for a *Lie group* the above relation between the parameters is assumed to be an analytic function [7–11]. Just as a reminder, the analytic property means that the relation between the parameters is expressed as a convergence of a positive power series of terms. Such analytic characteristics define a very special class of groups of transformations, which is called a Lie group of transformations.

One of the most relevant properties of a Lie group is that it makes it possible to derive the properties of the Lie group from a much simpler linear algebra, known as the *Lie algebra associated with a Lie group*. Using such Lie algebra, a Lie group can be seen as an analytic manifold, similar to a differentiable manifold, but with the abovementioned analytic relationship between the parameters. However, there is a difference: In a Lie group, the relation between parameters is analytic instead of being just differentiable. This makes a difference that is frequently neglected. We will return to this relevant distinction, including the differences on the associated geometries in Chap. 7.

The above mentioned analytic property of the Lie group means that all representations are isomorphic to the tangent space to the unit element of the same Lie group. Such a property enormously simplifies the representations of a Lie group because they can be obtained from the much simpler representations of the respective Lie algebra. The importance of being a Lie group in gauge field theory cannot be neglected: It was found that the above described Maxwell's equations can be generalized by the nonlinear Yang-Mills equations, retaining the Lie group characteristics. In fact, the Yang-Mills equations represent three among the four known fundamental forces of nature. The analytic property may also explain why all gauge fields are renormalizable, as discussed in Chap. 5.

1.2 Excerpts from Noether's Theorem

Most of the textbooks referring to Noether's theorem exemplify its application to the cases of the conservations of energy and momentum in classical mechanics, frequently omitting the applications to gauge symmetries. Due to the relevance of that theorem to the remaining of this book, here we detail the derivation of Noether's theorem for global and local gauge symmetries. Here we follow the rare but detailed demonstration of Noether's theorem from [12].

1.2.1 Noether's Theorem for Global Gauge Symmetries

Consider a generic field denoted by $\psi(x)$ with N-parameter symmetry, defined by a Lagrangian functional $\mathscr{L}(\psi, \psi_{,\mu})$, which is invariant under a unitary N-parameter gauge symmetry $SU(N)$, where the parameters θ are not dependent on the coordinates. Then, there exist N quantities $\mathscr{N}_a{}^{\lambda}$ which are conserved in the sense that their divergence vanishes $\mathscr{N}^{\lambda}_{a,\lambda} = 0$.

By definition the *total variation of the action* $\delta_T A$ of the field ψ, produced by a symmetry transformation, vanishes

$$\delta_T A = \int_{\Omega} \left(\frac{\partial \mathscr{L}}{\partial \psi} \delta_T \psi - \frac{\partial \mathscr{L}}{\partial \psi_{,\lambda}} \delta_T \psi_{,\lambda} \right) dv = 0. \tag{1.13}$$

Applying the Euler-Lagrange equations for ψ:

$$\frac{\partial \mathscr{L}}{\partial \psi} - \frac{\partial}{\partial x^{\lambda}} \left(\frac{\partial \mathscr{L}}{\partial \psi_{,\lambda}} \right) = 0,$$

it follows that the total variation of the Lagrangian also vanishes

$$\delta_T \mathscr{L} = \frac{\partial \mathscr{L}}{\partial \psi} \delta_T \psi + \frac{\partial \mathscr{L}}{\partial \psi_{,\lambda}} \delta_T \psi_{,\lambda} = 0.$$

In the case of a global gauge transformation, the parameters θ^a do not depend on the coordinates, so that the total variation of the field $\delta_T \psi$ can be seen to reduce to the variation of the parameters $\delta \theta^a$

$$\delta_T \psi = \sum \mathscr{F}_a(x, \psi) \delta_T \theta^a,$$

where the coefficients $\mathscr{F}_a(x, \psi)$ depend explicitly only on the field ψ and the coordinates, but they do not depend on θ^a. Consequently, the variation of the action (1.13) gives

$$\delta_T A = \theta^a \int_{\Omega} \frac{\partial}{\partial x^{\lambda}} \left(\frac{\partial \mathscr{L}}{\partial \psi_{,\lambda}} \mathscr{F}_a \right) dv = 0.$$

Defining the quantity

$$\mathscr{N}^{\lambda}{}_a = \frac{\partial \mathscr{L}}{\partial \psi_{,\lambda}} \mathscr{F}_a, \tag{1.14}$$

it follows that for a closed region Ω where the variables ψ, their derivatives $\psi_{,\lambda}$, and the Lagrangian $\mathscr{L}$ are continuous functions, we obtain a vanishing divergence

$$\int_{\Omega} \mathscr{N}^{\mu}{}_{a,\mu} dv = 0. \tag{1.15}$$

Therefore, in the boundary $\partial\Omega$ of Ω, with an outside oriented normal vector dS_n, we may apply the divergence theorem

$$\int_{\partial\Omega} < \mathscr{N}_a, dS_n >= 0,$$

leading to the conclusion that there exist N conserved quantities $\mathscr{N}_a$ in the sense of their vanishing divergences.

1.2.2 Noether's Theorem for Local Gauge Symmetries

Consider again a generic field denoted by $\psi(x)$ with an N-parameter local symmetry group, defined by a Lagrangian functional $\mathscr{L}(\psi(x), \psi(x)_{,\mu})$, which is invariant under a unitary N-parameter gauge symmetry $SU(N)$, where the parameters $\theta(x)$ are functions of the space-time coordinates. Then, there exist N quantities

$$\mathscr{N}^{\lambda}{}_{a} = \frac{\partial \mathscr{L}}{\partial \psi_{,\lambda}} \mathscr{F}_a,$$

which are conserved with respect to the gauge covariant derivative

$$D_{\lambda} \mathscr{N}^{\lambda} = 0,$$

where $D_{\lambda} = I\partial_{\lambda} + A_{\lambda}$, I denotes the $N \times N$ identity matrix, and A_{λ} are matrices defined in the adjoint representation of the gauge Lie algebras.

As in the previous case, we start with the total variation of the action which like (1.13) also vanishes

$$\delta_T A = \int_{\Omega} \left(\frac{\partial}{\partial x^{\lambda}} \left(\frac{\partial \mathscr{L}}{\partial \psi_{,\lambda}} \right) \delta_T \psi + \frac{\partial \mathscr{L}}{\partial \psi_{,\lambda}} \delta_T \psi_{,\lambda} \right) dv = 0.$$

Now we have to consider that the parameters also depend on the coordinates, so that the total variation of the gauge field ψ and its derivatives depend also on the variation of the parameters θ^a.

$$\delta_T \psi = \mathscr{F}_a(x, \psi)\theta^a \quad \text{and} \quad \delta_T \psi_{,\lambda} = \frac{\partial}{\partial x^{\lambda}} (\mathscr{F}_a(x, \psi)\theta^a).$$

As before, after applying the Euler-Lagrange equations, the vanishing total variation of the action integral is

$$\delta_T A = \int_\Omega \frac{\partial}{\partial x^\lambda}\left(\frac{\partial \mathcal{L}}{\partial \psi_{,\lambda}}\delta_T \psi)\right) dv = \int_\Omega \frac{\partial}{\partial x^\lambda}\left(\frac{\partial \mathcal{L}}{\partial \psi_{,\lambda}}\mathscr{F}_a \delta\theta^a\right) dv = 0.$$

Note however that differently from the previous case, we cannot remove the term $\delta\theta^a$ from the last bracket because $\delta\theta$ depends on the coordinates x_μ. To remedy this, Noether expanded the indicated derivatives obtaining

$$\int_\Omega \left[\frac{\partial}{\partial x^\lambda}\left(\frac{\partial \mathcal{L}}{\partial \psi_{,\lambda}}\mathscr{F}_a\right)\delta\theta^a + \left(\frac{\partial \mathcal{L}}{\partial \psi_{,\lambda}}\mathscr{F}_a\right)\frac{\partial \delta\theta^a}{\partial x^\lambda}\right] dv = 0. \tag{1.16}$$

As we see, the term within brackets in the last expression is not a divergence, so that we do not obtain a conserved quantity. To solve this problem, Noether added and subtracted new terms within the brackets as shown below:

$$\left[\frac{\partial}{\partial x^\lambda}\left(\frac{\partial \mathcal{L}}{\partial \psi_{,\lambda}}\mathscr{F}_a\right)\delta\theta^a + \left(\frac{\partial \mathcal{L}}{\partial \psi_{,\lambda}}\mathscr{F}_a\right)\frac{\delta\theta^a}{\partial x^\lambda} + \underbrace{\frac{\partial \mathcal{L}}{\partial \psi_{,\lambda}}\mathscr{F}_a A^a_{\lambda n}\delta\theta^b}_{\text{added}} - \underbrace{\frac{\partial \mathcal{L}}{\partial \psi_{,\lambda}}\mathscr{F}_a A^a_{\lambda b}\delta\theta^b}_{\text{subtracted}}\right], \tag{1.17}$$

where the coefficients $A^a_{\lambda b}$ are defined by the additional condition

$$\sum_a \mathscr{F}_a A^a_{\lambda\,b}\delta\theta^b = \sum_a \mathscr{F}_a \frac{\partial \delta\theta^a}{\partial x^\lambda}. \tag{1.18}$$

Now, imposing that the second and fourth terms in (1.17) cancel each other:

$$\frac{\partial \mathcal{L}}{\partial \psi_{,\lambda}}\mathscr{F}_a\left(\frac{\delta\theta^a}{\delta x^\lambda} - A^a_{\lambda b}\delta\theta^b\right) = 0, \tag{1.19}$$

so that (1.16) reduces to,

$$\int_\Omega \frac{\partial}{\partial x^\lambda}\left(\frac{\partial \mathcal{L}}{\partial \psi_{,\lambda}}\mathscr{F}_a + \frac{\partial \mathcal{L}}{\partial \psi_{,\lambda}}\mathscr{F}_b A^b_{\lambda\,a}\right)\delta\theta^a dv = 0.$$

where we note the relevant fact that the term $\delta\theta^a$ in the previous expression was removed out of the derivative. To end, Noether defined the "gauge covariant derivative"

$$D^a_{b\lambda} = \delta^a_b \partial_{,\lambda} + A^a_{\lambda b}.$$

Denoting

$$\mathcal{N}^{\lambda}{}_{a} = \frac{\partial \mathcal{L}}{\partial \psi_{,\lambda}} \mathcal{F}_{a},$$

the expression (1.16) can be written as

$$\int_{\Omega} \left(D^{a}_{b\lambda} \mathcal{N}^{\lambda}{}_{a} \right) \delta\theta^{b} dv = 0.$$

Assuming that $\mathcal{N}^{\lambda}{}_{a}$ is differentiable in the integration region Ω, Noether obtained a *generalization of the classic divergence theorem with respect to the gauge covariant derivative*

$$\left(D^{a}_{\lambda\ b} \mathcal{N}^{\lambda}{}_{a} \right) \delta\theta^{b} = 0. \tag{1.20}$$

Since $\delta\theta^{a}$ are linearly independent parameters, after summing up all the internal indices $a, b, \ldots$, we obtain Noether's generalized expression for divergence theorem in the space-time.

$$D_{\lambda} \mathcal{N}^{\lambda} = 0, \tag{1.21}$$

where D_{λ} are the space-time components of gauge covariant $N \times N$ matrix operator. Denoting by I the $N \times N$ identity matrix, we obtain

$$D_{\lambda} = I \partial_{\lambda} + A_{\lambda}. \tag{1.22}$$

Note that for each fixed value of the coordinate index λ, $D^{a}_{\lambda\ b}$ has two indices a, b so that D_{λ} is an $N \times N$ matrix, whose components are defined in the space of the Lie algebra of the gauge group $SU(N)$. In other words, *D_{λ} is a matrix whose components are defined in the Lie algebra of the gauge group $SU(N)$. Thus, for each value of λ these matrices define a particular representation of the gauge group, whose representation space is the space of the Lie algebra itself. This is called the adjoint representation of the gauge group.*[6]

1.3 The Standard Model

The simplest gauge field is given by Maxwell's electromagnetic force, whose gauge symmetry has only one local (coordinate dependent) parameter $\theta(x)$. Such a gauge

[6] More specifically, if we denote by $\{X_{a}\}$ a base of a Lie algebra of a Lie group, then the *adjoint representation* of that group is a particular representation of that group, using the space of its Lie algebra as the space of representation, so that $D(X_{a})X_{b} = f^{c}_{ab} X_{c} = [X_{a}, X_{b}]$.

group is isomorphic[7] to the group $SO(2)$ of rotation on a circle, so that each quantity $A^{a}_{\lambda\ b}$ in Noether's theorem is just the relativistic electromagnetic potential vector, with real components: A_λ playing the role of the gauge affine connection of the affine geometry of the classical electromagnetism (we shall detail the notion of affine geometry later on in Chap. 4).

The next known gauge field is called the *weak nuclear force* associated with $SU(2)$ isomorphic to the $SO(3)$ group of rotations on a spherical internal surface proposed by E. Fermi in 1933. Later on C. N. Yang and T. D. Lee noted that the weak interaction could not be stable (or to use the language of quantum field theory, the $SU(2)$ symmetry could be broken) which would imply in the possible emergence of unknown particles. The problem of stability was solved in the 1960s when Sheldon Glashow, Abdus Salam, and Steven Weinberg unified the electromagnetic force and the weak symmetry in the form of the Cartesian product of the respective Lie algebras $U(1) \times SU(2)$, which became known as the electroweak symmetry.

Since the product symmetry $U(1) \times SU(2)$ where the factor groups acts, respectively, in a one-dimensional space (a circle S_1) and a two-dimensional spherical surface S_2, then the product symmetry acts on the three-dimensional internal space defined at the quantum scale. Such product symmetry was confirmed in 1983, with the experimental confirmation of the predicted results. The electroweak field became the second example of unification of fundamental gauge forces after the electromagnetic unification. It produced a new fundamental force of nature called the *electroweak force* [13, 14].

The third and last gauge symmetry becomes evident when it was noted that the nucleus of an atom was composed by protons and neutrons, bound together what became known as the strong nuclear force, as illustrate in the Fig. 1.2 [15–18].

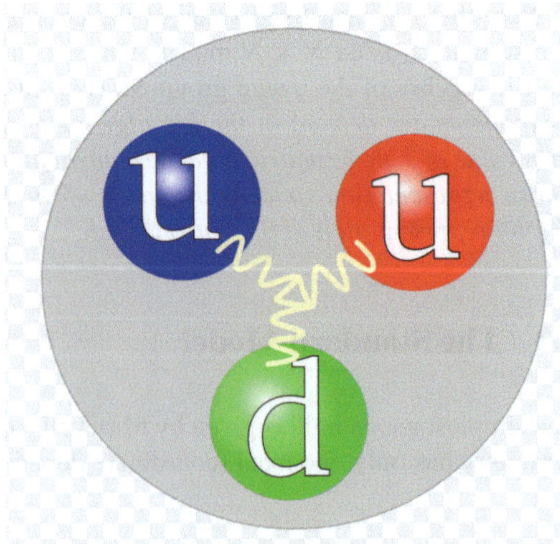

Fig. 1.2 Two up and one down quarks bound by strong nuclear forces in the nucleus of an atom

[7] From its definition it follows that all Lie groups with the same number of parameters are isomorphic to each other.

Note that differently from the other gauge groups, the $SU(3)$ group is not related to a rotational group. Actually, $SU(3)$ is the smallest subgroup of the class G_2 of *exceptional Lie groups*, which is also isomorphic to the group of automorphisms of the octonion algebra.

In 1954 C. N. Yang and R. Mills found that all three gauge fields would satisfy the same set of equations. These are similar to the mathematical expressions of Maxwell's equations, except for the fact that the gauge potentials are described in the respective adjoint representations of the Lie groups $U(1)$, $SU(2)$, and $SU(3)$:

$$D^{\mu} F_{\mu\nu} = 4\pi J_{\nu}, \quad D^{\mu} F^{*}_{\mu\nu} = 0, \tag{1.23}$$

which today are referred to as the Yang-Mills equations, derived from the Yang-Mills Lagrangian

$$\mathscr{L}_{Yang-Mills} = -\frac{1}{4} tr\, F_{\mu\nu} F^{\mu\nu}. \tag{1.24}$$

This Lagrangian has the same appearance of the electromagnetic Lagrangian functional, except for the presence of the trace in (1.24), required to make it the simplest scalar invariant Lagrangian out of the matrix structure of the gauge forces $F_{\mu\nu}$ [19].

Another important aspect to keep in mind is that the Yang-Mills equations hold only in four-dimensional space-times. One simple way to see this is to use the language of differential forms to write the Yang-Mills tensor equation (1.23) as the 2-form $F = \frac{1}{2} F_{\mu\nu} dx^{\mu} \wedge dx^{\nu}$, and the current vector is one form $J = J_{\mu} dx^{\mu}$. Using such language we find that the covariant derivative $D = D_{\mu} dx^{\mu}$, which acts on the 2-form F through the exterior product [20]

$$D \wedge F = 4\pi J, \qquad D \wedge F^{*} = 0. \tag{1.25}$$

Such a result evidences a special geometrical property of the gauge fields that cannot be generalized to arbitrary dimensions: The manifold structure of the Yang-Mills equations implies a limitation on the structure of mathematics that cannot be extended beyond four dimensions.

The Standard Model of Fundamental Interactions represents what is known within the presently available theoretical and experimental resources available. It can be summarized by the table of fundamental particles and forces that act as the most fundamental objects in nature (Fig. 1.3).

Note that the gravitational force is not represented in the above table, in spite of the general belief that it is a fundamental force of nature. As such it should belong to the Standard Model. However, up to the present there is no experimental evidence indicating that the gravitational field associated with Einstein's classical theory of General Relativity can fit in the Standard Model. This is in part due to the presence of the Newtonian gravitational constant in Einstein's equations, imposing a strong

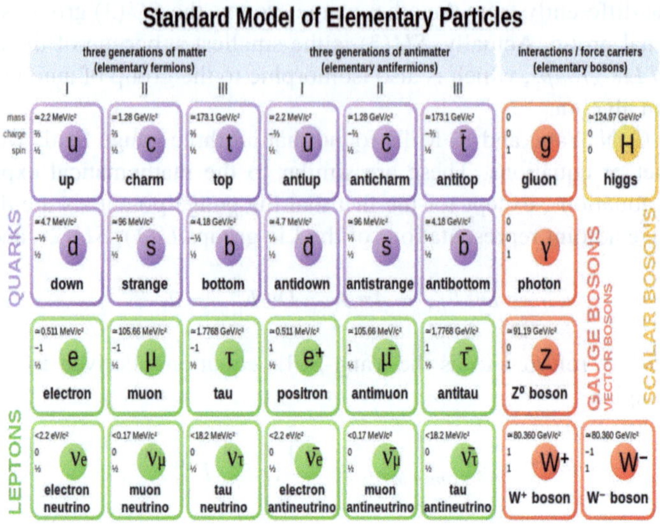

Fig. 1.3 The present constitution of the Standard Model of the Fundamental Interactions does not include the quantum of the gravitational force

hierarchy between gravitation and the three gauge interactions. At the end we will see that in fact gravitation may be part of the whole structure of the Standard Model, into an extended geometrical structure that generalizes four-dimensional manifold theory [21, 22].

Chapter 2
Gravitation and Cosmology

Abstract While in Newtonian gravitation the presence of gravitation is measured by the gravitational pull between two lumps of atoms, composing what we call ordinary matter, in Einstein's General Relativity, the presence of gravitation is associated with the curvature of the space-time, regardless of the nature of the gravitational source. However, the recent observations from infrared and ultraviolet astronomy show that the contribution of ordinary matter to the universe accounts for only a small fraction.

2.1 General Relativity

Here we briefly review Einstein's gravitational theory as a geometric generalization of Newtonian Gravitation. We follow some details of Penrose's Adams prize [6]. Questions concerning quantum gravity will be studied separately in the next chapter.

We start with Newton's gravitational law, describing the gravitational force between two ordinary lumps of masses separated by a distance. According to Newton, the gravitational force is given by

$$F = \mathbf{G}\frac{mm'}{r^2}, \tag{2.1}$$

where **G** is the Newtonian gravitational constant. The above equation is invariant under the group of transformations called the generalized Galileo group of symmetries composed of the following changes of coordinates and Newton's universal time:

$$\begin{cases} x'^i &= a^i{}_j(t)\, x^j + c^i(t) + d^i \\ t' &= at + b \\ \phi' &= \phi + \displaystyle\sum_{i=1}^{3} \frac{d^2a^i}{dt^2}\, x^i \end{cases} \tag{2.2}$$

where the parameter t denotes the Newtonian absolute time.

M. D. Maia, E. M. Monte, *Geometry at the Quantum Scale*,
https://doi.org/10.1007/978-3-031-61197-1_2

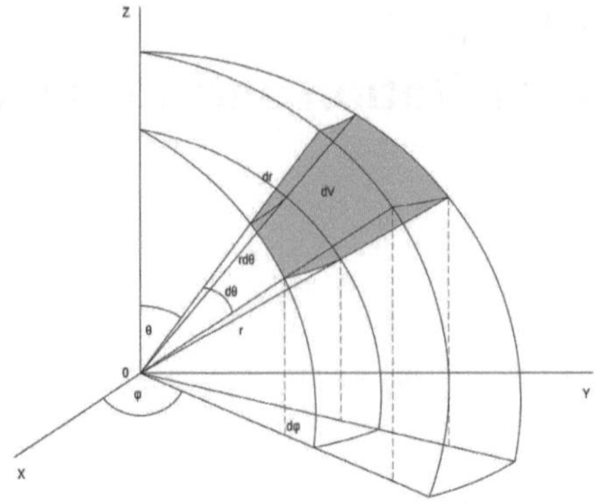

Fig. 2.1 The solid angle in spherical coordinates

In order to express Newton's gravitational equation (2.1) as a *differential equation*, consider the Newtonian gravitational field generated by a uniform and spherically symmetric distribution of matter centered at the origin $r = 0$ of a spherical coordinate system (r, θ, φ) with density ρ. Take $m = 1$ as an initial mass located at the origin $r = 0$ and m' as an infinitesimal mass uniformly distributed in a solid angle section of a spherical shell distant r as shown in the Fig. 2.1.

We may express the mass of the small conic section with volume dv (the shaded volume in the figure) as $m' = \rho dv = \rho dr sin\theta d\theta d\varphi$. Next, integrating in θ from π to $-\pi$ and φ from 0 to 2π, we obtain from (2.1) that the infinitesimal force oriented toward the center of the sphere is given by

$$dF = -4\pi \mathbf{G}\rho dr. \tag{2.3}$$

Using the conventional orientation of the surface of a 3-sphere to be pointing *toward the outside, assuming that the density ρ remains constant in each spherical shell*, and finally considering that the Newtonian gravitational force is conservative (i.e., derived from a potential $F = -\nabla\phi$), we obtain the infinitesimal force $dF = -\nabla^2\phi$. Comparing this last expression with (2.3), we obtain the Newtonian gravitational equation written as the well-known (Poisson) differential equation

$$\nabla^2\phi = 4\pi\mathbf{G}\rho. \tag{2.4}$$

The value of **G** is obtained from the gravitational attraction force between two small masses separated by a small distance (as compared with the human scale), using a torsion balance. More recently, the Casimir effect has been used in place of the torsion balance to obtain a more precise measurement when it was then possible to confirm the value of **G** down to the distance of only four microns, 4×10^{-3} mm [23]. The current value of **G** has been verified to be

$$\mathbf{G} = 6 : 67430.10^{-11}\frac{m^3}{kgs^2}. \tag{2.5}$$

On the astronomical scale, the evaluation of **G** was estimated by planetary motions, using the astronomical instruments available at the beginning of the twentieth century. These measurements led to the assumption that **G** is one of the celebrated universal constants of nature along with the reduced Planck constant $\hbar$ and the speed of light c.

The Newtonian gravitational theory remained the standard explanation for the gravitational force up to 1915, when Einstein presented his new geometric theory of gravitation. Einstein's approach to gravitation was based on Riemann's geometry defined on manifolds. A detailed reference on *Riemannian geometry*, using coordinates, can be found in Eisenhart's classic book Riemannian Geometry [24]. For those who prefer a coordinate-free approach, there are many modern references, among which we may cite [25, 26].

Riemannian geometry extends the Euclidean positive metric geometry to differentiable manifolds, in which a point p is described by any number of arbitrary coordinates $(x^1, x^2, x^3, \ldots x^n)$, such that the distance between any two infinitesimally close points is given by the quantity (the metric)

$$ds^2 = g_{\mu\nu} dx^\mu dx^\nu,$$

where $g_{\mu\nu}$ denotes the components of the metric tensor **g** in a given coordinate system.

A fundamental assumption of Riemann's geometry is that the distance between two points must be independent of the coordinates that specify the points. This implies that $g_{\mu\nu}$ must transform in a specific way described in Riemannian geometry as

$$g'_{\mu\nu} = \frac{\partial x^\rho}{\partial x^\mu} \frac{\partial x^\sigma}{\partial x^\nu} g_{\rho\sigma}.$$

The independence of the metric from a choice of coordinates may also be expressed in terms of the *covariant derivative* of the metric tensor denoted by $\nabla_\rho g_{\mu\nu} = 0$. To eliminate misunderstandings, we generally use a semicolon to denote the covariant derivative, except in some classic books (like, e.g., in Eisenhart's book).

$$\nabla_\rho g_{\mu\nu} = 0, \quad \text{or equivalently} \quad g_{\mu\nu;\rho} = 0,$$

while the ordinary partial derivative remains written with a simple colon $g_{\mu\nu,\rho}$. From the point of view of geometry the above expression is equivalent to say that the observer may walk around the manifold carrying a local scale as a tool to locally measure distances, literary justifying the meaning of the word "geometry" borrowed from land surveying. Apparently, Riemannian geometry was originally conceived as the geometry for the local observers' space. This has not prevented the applications of Riemannian geometry to distant phenomena, like in cosmology and quantum scales, where the observers cannot reach, *provided the method of observations is adapted to distant observations*. This is not a real problem because the observers may opt to define *an appropriate affine connection* so that the observer may compare distant objects using the local metric scale as gauge standard. The history of development of Riemannian geometry tells a different story. In 1901 Tullio Levi-Civita defined the affine connection of Riemannian geometry as being the same Christoffel symbols associated with the metric [27]. However, as every geometer knows, the *Christoffel symbols* of the first and second species, respectively, are obtained by the cyclic permutation of the indices μ, ν, ρ in the derivatives of the metric tensor:

$$\Gamma_{\mu\nu\rho} = \frac{1}{2}(g_{\mu\nu,\rho} + g_{\mu\rho,\nu} - g_{\nu\rho,\mu}) \quad \text{and} \quad \Gamma^{\rho}_{\mu\nu} = g^{\rho\sigma}\Gamma_{\mu\nu\sigma}, \tag{2.6}$$

and here lies the problem for observational physics based on Riemannian geometry: Different observers will not agree on what they are looking at. On the other hand, the Levi-Civita affine connection may be combined in a special way to generate the *Riemann curvature tensor* given by

$$R_{\mu\nu\rho\sigma} = (\Gamma_{\nu\rho\mu,\sigma} - \Gamma_{\nu\sigma\mu,\rho}) + (\Gamma^{\tau}_{\nu\rho}\Gamma_{\mu\sigma\tau} - \Gamma^{\tau}_{\mu\rho}\Gamma_{\nu\sigma\tau}). \tag{2.7}$$

From these components we obtain the trace $R_{\mu\nu} = g^{\mu\rho}R_{\mu\nu\rho\sigma}$ which is called the Ricci tensor, and its trace $R = g^{\mu\nu}R_{\mu\nu}$ which is called the Ricci scalar. Therefore, as Einstein found out, it is possible to obtain a generalization of the Newtonian gravitational field, assuming that the four-dimensional classical space-time has a Riemannian geometry defined by the metric tensor *with the Levi-Civita affine connection*, such that the local observables are given by the curvature tensor (2.7) of the space-time.

We may follow Einstein's reasoning with a simple generalization of the Newtonian gravitational field, by defining a four-dimensional Riemannian space-time whose metric is associated with the Newtonian gravitational potential as

$$(g_{\mu\nu}) = \begin{pmatrix} 1 & 0 & 0 & 0 \\ 0 & 1 & 0 & 0 \\ 0 & 0 & 1 & 0 \\ 0 & 0 & 0 & -\phi \end{pmatrix}, \tag{2.8}$$

where ϕ is the same Newtonian gravitational potential that appears in Poisson's equation (2.4). The minus sign in ϕ follows from the assumption that such space-time geometry will have the same signature of the Minkowski space-time in the absence of gravitation. Likewise, the source of gravitation is specified by the density of matter ρ placed in the right-hand side of (2.4):

$$(T_{\mu\nu}) = \begin{pmatrix} 1 & 0 & 0 & 0 \\ 0 & 1 & 0 & 0 \\ 0 & 0 & 1 & 0 \\ 0 & 0 & 0 & -\rho \end{pmatrix}. \tag{2.9}$$

With this simple example of a four-dimensional geometry, calculating the Christoffel symbols from (2.8), we obtain the Riemann tensor and the Ricci tensor. The latter has a single nontrivial component R_{44}. Comparing this with the corresponding T_{44} component, we obtain the Poisson equation (2.4), written as

$$R_{44} = 4\pi \mathbf{G} T_{44}.$$

This expression suggested that the Poisson equation is just the 4×4 component of a new and more general equation for the gravitational field written in generic coordinates as

$$R_{\mu\nu} = 8\pi \mathbf{G} T_{\mu\nu}, \tag{2.10}$$

where now $R_{\mu\nu}$ is taken to represent the Ricci curvature of a new and more general four-dimensional space-time, and $T_{\mu\nu}$ represents the components of a new and more general source term of the gravitational field.

Note that $\mathbf{G}$ in the equations (2.10) is the same. Newtonian gravitational constant, so that in principle the source of the gravitational field in (2.10) is the same as the ordinary matter in Newton's gravitation.

Soon Einstein perceived that the above equation is not consistent because the right-hand side represents the energy momentum tensor, and as such it must be conserved in the same sense of Noether's theorem but now generalized for a Riemannian geometry, and in that principle it should apply to coordinate system. Therefore *the contracted covariant derivative of $T_{\mu\nu}$ must be zero*: $T^{\mu\nu}{}_{;\nu} = 0$. However, the covariant divergence of the left-hand side of the equation $\nabla_\nu R^{\mu\nu} = 0$ does not necessarily vanish. To solve this problem, Einstein defined a new tensor whose covariant derivative also vanishes. This new tensor is known today as the *Einstein tensor*, satisfying the requirement $G^{\mu\nu}{}_{;\nu} = 0$. Therefore Einstein replaced Eq. (2.10) by

$$G_{\mu\nu} \stackrel{def}{=} R_{\mu\nu} - \frac{1}{2} R g_{\mu\nu}. \tag{2.11}$$

With such provision, Einstein concluded that

$$G_{\mu\nu} = R_{\mu\nu} - \frac{1}{2} R g_{\mu\nu} = 8\pi \mathbf{G} T_{\mu\nu}, \tag{2.12}$$

which is the present form of Einstein's equations (except for the inclusion of cosmological constant which will be discussed below within the context of cosmology).

Note that (2.12) is a system composed of 13 differential equations, one of them being nonlinear, whose solution defines the components of the metric tensor $g_{\mu\nu}$ of the classic space-time, representing the gravitational potential of Einstein's General Relativity theory [28].

Einstein's gravitational theory defined by (2.12) was soon subjected to three basic experiments, namely the advance of Mercury's perihelion precession, the bending of light by gravitation, and the gravitational redshift [5], which were predicted from Einstein's calculation. These results led to an almost immediate acceptance of General Relativity as the new theory of gravitation holding such status to the present day.

As in any new fundamental theory, its observable structures must be defined. In the case of Einstein's gravitation, these observables are determined by the measurement of the local curvature of the space-time as seen by an observer belonging to the space-time itself. This is not too obvious, except if we follow the above developments of Einstein's equations (2.12), where we note the present gravitational constant **G** in the right-hand side.

This brings up a geometrical problem of how we (the observers) may measure the curvature of the space-time manifold, if the observer lives in the space-time itself? Actually this was already defined in Riemannian geometry: Using the index symmetries of the Riemann tensor, we may construct a basis of eigenvectors $X^{\mu\nu}$ of the Riemann tensor in the space-time, such that $X^{\mu\nu} = -X^{\nu\mu}$. In such basis, the curvature components are given by the respective eigenvalues $\lambda_{\mu\nu}$ solutions of the eigenvalues equation

$$R_{\mu\nu\rho\sigma}X^{\rho\sigma} = \lambda_{\mu\nu}X^{\rho\sigma}, \qquad \text{no sum on } \rho \text{ and } \sigma. \tag{2.13}$$

These eigenvalues are always measured with respect to the zero local curvature which represents the flat tangent Minkowski space-time. By counting the indices from 1 to 4 and taking into account the repeated values (index symmetries) in the eigenvalue equations, we find that only five nonzero and independent eigenvalues remain to completely determine the local curvature of the space-time, or if you like, to determine the relativistic gravitational force at the point where these values are calculated [29]. Equivalently, the gravitational field of Einstein's gravitation is said to have *five degrees of freedom* or $dof = 5$.

The above curvature of the space-time refers to curvature of the classical space-time of the observers. For the space-time defined in the quantum domain, we use a different designation proper of quantum theory. The spin-statistic theorem of quantum field theory relates the degrees of freedom in terms of a quantum parameter, to the quantum spin as [30]

$$s = \frac{dof - 1}{2} = 2.$$

Therefore, the five degrees of freedom of the quantum space-time correspond to a quantum spin-2 field or particle. In other words, we say that Einstein's

Type I $\longleftrightarrow$ Four simple independent eigenvectors
Type II $\longleftrightarrow$ One double + Two simple independent eigenvectors
Type III $\longleftrightarrow$ One triple + One simple independent eigenvectors
Type IV (or N) $\longleftrightarrow$ One quadruple independent eigenvector
Type V (or D) $\longleftrightarrow$ Two double independent eigenvectors
Type O $\longleftrightarrow$ all eigenvectors vanish

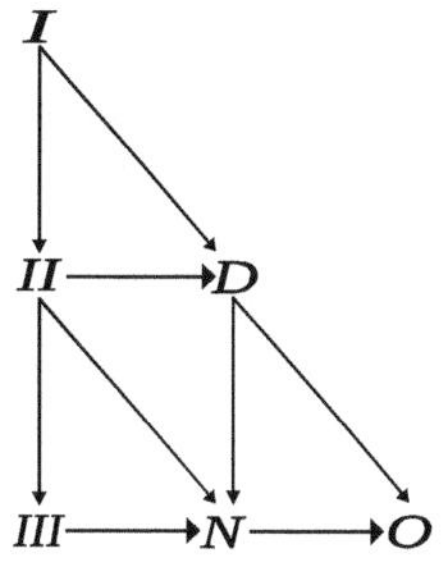

Fig. 2.2 The Petrov classification of the space-time curvature

gravitation corresponds to a spin-2 quantum particle. This is an intrinsic geometric characteristic that is independent of the observer or how the observer measures the curvature. If less than four degrees of freedom are measured, then the curvature of the quantum space-time geometry or equivalently the quantum spin of the corresponding quantum gravity is also incomplete.

The number $(dof - 1) = 4$ in the above expression denotes the minimal number of test probes required to measure the gravitational field in the classical space-time. The nontrivial eigenvalues of the Riemann tensor are classified by the Petrov Classification shown in Fig. 2.2 [31, 32]. The maximum number four of independent eigenvectors makes it possible to construct a special coordinate basis in space-time, called the Newman-Penrose basis, in which the curvature of the space-time is expressed in terms of the above Petrov Types [33].

The identification of the gravitational force with the curvature of the space-time implies that the definition of the observers is not the same as in the Minkowski space-time of Special Relativity, which are mapped to one another by the Poincaré symmetry. Instead, Einstein postulated that all observers are mapped to one another by the *group of diffeomorphisms* of the space-time, which is the basis of the principle of general covariance stating that *all observers are equally good*. Although this may sound as a political statement, in fact, it has nothing to do with social choices. As we will detail in the next chapter, the general covariance is not a principle but rather an *inevitable* condition which is built in the Einstein-Hilbert Action Integral.

2.2 Cosmology

In the present Standard Model of the universe, its material content is assumed to be made by ordinary matter. This means that only the atomic structure is taken into consideration, ignoring the nuclear forces acting inside the nucleus of the atoms. This conclusion stems from the presence of the Newtonian coupling constant

G in Einstein's equations, which translates the Newtonian gravitational attraction between two ordinary macroscopic lumps of matter, ignoring the nuclear forces that compose the Standard Model. Actually, up to the first quarter of the twentieth century, cosmology was still (and still is) mostly seen as a phenomenon of classical physics.

The present standard model of the universe emerged by 1922, when a solution of Einstein's equations was described by A. Friedman using the cosmological principle, stating that the universe is homogeneously and isotropically composed of ordinary matter [34]. This homogeneity principle makes it possible to consider an observer sitting at the center of a sphere at any point of the universe (like in Fig. 2.1), expanding equally in all directions, due to the isotropic assumption. In particular, taking a solid angle between $\theta + d\theta$, $\phi + d\phi$ separated by a radial distance between $r = r = dr$, such a universe may be described as a perfect fluid in expansion, with velocity v_μ. Then the observer at the origin of this spherical coordinate system may describe a small section of the universe located at a distance r from the observer given by what is called the line element of the universe, describing the distance between two close points in the Universe written as (here we follow mostly [35])

$$ds^2 = g_{\mu\nu} dx^\mu dx^\nu = a^2(t) \left[\frac{dr^2}{1 - kr^2} + (r^2 d\theta^2 + r^2 \sin^2 \theta d\phi^2) \right] - dt^2, \quad (2.14)$$

where r is the radial coordinate, $a(t)$ is called the *expansion factor of the universe*, and k is a *topological factor* specifying shape of the three-dimensional space-like hypersurfaces between in the universe (as shown in Fig. 2.3).

On the other hand, the material source of such isotropic and homogeneous universe is assumed to be given by the energy-momentum tensor of a perfect fluid, whose (space-like) components are given by the components

$$T_{ij} = (p + \rho) v_i v_j + p g_{ij} \text{ and } T_{44}, = 0,$$

where ρ denotes the density of the perfect fluid and p stands for its pressure. Such a tensor is conserved with respect to the covariant derivative defined by the metric (2.14)). Finally, the state equation for a perfect fluid is given by $p = (\gamma - 1)\rho$, with $(1 \geq \gamma \leq 2)$, telling that the density of the perfect fluid varies with the expansion of the universe as

$$\rho(t) = M \left(\frac{a_0}{a(t)} \right)^{3\gamma}, \quad (2.15)$$

where a_0 is an initial value for the expansion factor $a(t)$, M is the mass density unit factor, and γ depends on the nature of the matter. For example, for a dust-like distribution of matter, then, $\gamma = 1$.

Replacing the above expressions in Einstein's equations, we obtain the Friedman equations for the presently acknowledged *Standard Model of the Universe* given by the solution of the equations

$$\begin{cases} \dot{a}^2 = \frac{8\pi \mathbf{G}}{3}\rho a^2(t), \\ \frac{d(\rho a^3(t))}{dt} = -3pa^2(t), \\ p = (\gamma - 1)\rho. \end{cases} \tag{2.16}$$

Along many years the Friedman model received substantial contributions by G. Lemaitre (with the hypothesis of the big bang), H.P. Robertson (with the hypothesis of homogeneity and isotropy), and Howard P. Robertson and A. G. Walker (on the uniqueness of the solution), leading to the acronym FLRW.

The hyperbolic characteristics of Einstein's equations play a significant role in cosmological observations, in the sense that the line element (2.14) implies that the universe expands as a time-like sequence of three-dimensional space-like hypersurfaces, characterizing a foliation of the universe by three-dimensional hypersurfaces. The factor k was introduced for generality, considering that those hypersurfaces may be spherical surfaces ($k = 1$), or they may be planes ($k = 0$), or they may be hyperboloids ($k = -1$), as shown in the Fig. 2.3:

However, the astronomical observations of three stars with the same redshift (meaning that they belong to the same three-dimensional hypersurface in a distant past) has been consistently found that the sum of the internal angles of the triangle formed by these stars is always 180°, indicating that those hypersurfaces are flat, which means $k = 0$. Whatever is the shape of these surfaces, these observations are telling that the expansion of the universe obeys the hyperbolic characteristic of Einstein's equations.

Radio astronomy changed dramatically in 1965, quite by accident: During the installation of a large radio telescope antenna operating at the microwaves frequency, the astronomers Allan Penzias and Robert Wilson found a persistent

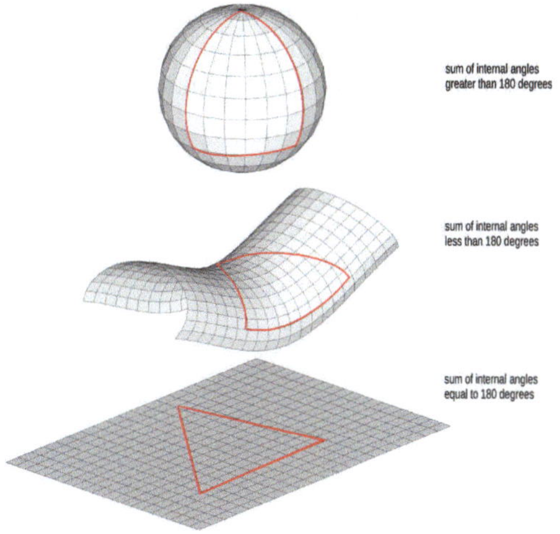

Fig. 2.3 Hypersurfaces with positive, zero, and negative curvatures

background noise in their receiver, which was independent of time and orientation of the antenna. After many attempts to eliminate possible sources for that background "noise," they concluded that it was due to an unpredicted presence of some kind of radiation of the universe at the microwave length, which was named the *Cosmic Microwave Background Radiation (CMBR)*. The FRLW model still applies, but due to such evidence, the parameter γ in the state equation of the perfect fluid (2.16) should be replaced by $\gamma = 4/3$ [35].

Still in the observational side, after 20 years of much effort G. Smoot managed to launch a satellite microwave antenna in 1989, orbiting around the Earth, called the Cosmic Background Explorer (COBE), to confirm and find a more detailed account of the CMBR. The result gave support to the evidence of the cosmic radiation, including to the big bang origin of the universe proposed by G. Lemaitre. Accordingly, immediately after the initial explosion, the universe was essentially composed of a "radiating energy" which produced the present universe. It is estimated that after the universe cooled down to about 3000°C, most of that radiation condensed into matter particles, and eventually it led to their agglutination of such matter in the form of atoms. The CMBR is the remnant of that initial radiation.

Interestingly, the COBE experiment also indicated the existence of a faint difference in the distribution of that radiation (see the Fig. 2.4), which motivated the construction of a new satellite antenna with more sensitive probes called the Wilkinson Microwave Anisotropy Probe (WMAP), launched in 2001. The result confirmed that indeed the cosmic radiation was not homogeneous, but it represented an anisotropic, distribution of patches that emit cosmic radiation and others that absorb such radiation (respectively, red shades and blue shades in the Fig. 2.4). The importance of the CMBR motivated the construction of a third satellite antenna, called the *Planck experiment* that was launched in 2009, confirming and giving more detail on the distribution of the CMBR anisotropy.

Indeed, the detailed analysis of the power spectrum of the CMBR brought a new insight on how the energy of the universe is distributed: Ordinary matter (composed of atoms) contributes to mere 4.9% of the total energy content of the universe. The remaining is unknown, forming what is called the dark sector of the universe, of which 26.8% is composed of an unknown substance called dark matter, and 68.3% of it is in the form of some unknown dark energy.

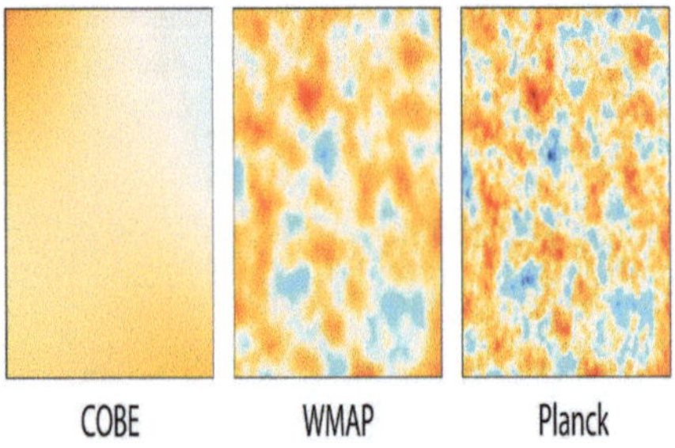

Fig. 2.4 Comparing the Cobe, WMAP, and Planck experiment resolutions

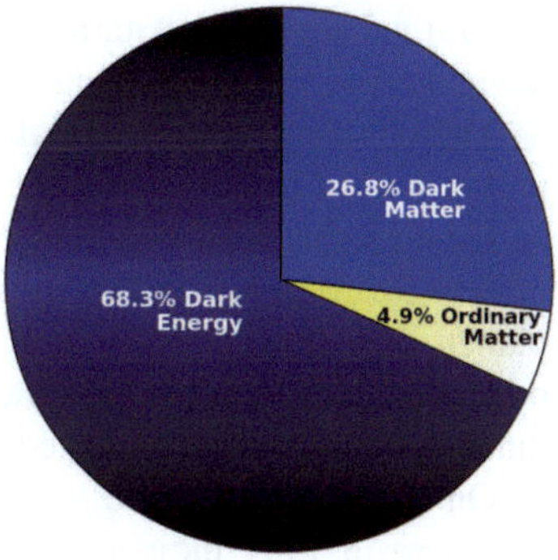

Fig. 2.5 Ordinary matter, dark matter, and dark energy in the universe

The CMBR experiments have dramatically impacted our current understanding about the universe: Prior to that, the common assumption was that the universe is composed essentially by ordinary matter. Since the Newtonian Gravitational constant **G** measures the gravitational pull between two lumps of ordinary matter, then the current estimates on the origin, age, size and composition of the universe, up to now, are affected by the presence of the Newtonian gravitational constant assumed to hold throughout the entire universe, assumed to be a universal constant. If so, then the most recent CMBR results tell that the composition of the universe corresponds to a mere 4.9% of the total energy content of the universe, as shown in Fig. 2.5.

The name dark matter appeared in 1933 when astronomical observations indicated that the velocity of the stars in the outermost arms of some spiral galaxies is too much high, about 200 km/sec, beyond the scape velocity of those stars. If we count only the visible masses as contributing to the gravitational pull of the galaxy, then the mentioned stars would have a higher velocity than it would be possible to be sustained by the gravitational field of the visible stars. The problem persists even with the supposition that large planets that do not emit light also contribute to the gravitational pull of the galaxy. Some possible alterations of the Newtonian law of gravitation were also proposed [36], but they proved to be insufficient to keep those stars in orbit. Consequently, we were led to the hypothetic existence of an unknown type of new matter, generically called "dark matter."

The rotation curves of stars tend to agree with the observations only near the nucleus of the spiral galaxies.

The velocities of rotation curves in the Fig. 2.6 refer to the rotation curves of stars in a spiral galaxy, simply because it is easier to astronomers to measure the rotation of the observed spiral galaxies with an attached rotation disk on their telescope's lenses. The continuous line (B) in Fig. 2.6 shows the observed rotation curve, while the dashed line (A) shows the theoretical rotation curve of stars as described by General Relativity. Note that the observed line is practically constant (sometimes referred to as flat rotation curves) as compared with the curve predicted by General Relativity. The two curves tend to agree only near the galactic center. The graph suggests that the contribution of ordinary visible matter of the gravitational pull in a spiral galaxy to the rotation curve is really insignificant, except at the nucleus of the galaxies. Are we missing something here or are we using the wrong parameter?

The observations show that if we take into consideration only the gravitational pull produced by ordinary light emitting (visible) matter, then those stars should not be there in their observed orbits [37, 38]. After many suggestions to explain dark matter, for example, by the possible inclusion of the gravitational pull from large planets and other possible non-visible sources (like black holes), the existence of other unknown sources of attractive gravitation, not composed of common visible matter or energy, was admitted as a fact without previous explanation.

On the other hand the observations of type I supernovae by the Hubble space telescope in 1997 indicated that the (observed) universe was also subjected to an *accelerated expansion*, as if subjected to some kind of negative source of energy. These measurements were confirmed by the WMAP and Planck satellites, indicating that 68.3% of the total energy of the universe is due to such dark energy. As we see, in Fig. 2.5 ordinary matter contained in the presently existing classic cosmological models only responds to less than 5% of the total energy content of the universe. As we can clearly see, we are missing an astonishing 95.1% of whatever composes the universe, leading to the conclusion that if Einstein's geometric gravitation is to play a relevant role in the observed universe, then other sources of gravitation must also be considered besides just ordinary atomic matter.

Interestingly, back in 1917 Einstein noted that in the derivation of his gravitational equations he had forgotten to add a mathematically admissible term proportional to the metric $\Lambda g_{\mu\nu}$, where Λ is called the cosmological constant, leading to the more general set of Einstein's equations, given by

$$R_{\mu\nu} - \frac{1}{2}Rg_{\mu\nu} + \Lambda g_{\mu\nu} = 8\pi \mathbf{G} T_{\mu\nu}. \tag{2.17}$$

In that same year W. de Sitter presented two vacuum ($T_{\mu\nu} = 0$) solutions of the above extended Einstein's equations with the cosmological constant Λ, one with $\Lambda > 0$ and the other with $\Lambda < 0$, called the de Sitter space-time and the anti-de Sitter space-time, respectively, representing space-times with constant curvature, whose metrics are invariant under the symmetry groups $SO(4+1)$ and $SO(3+2)$, respectively, both with ten parameters. These symmetry groups are isomorphic to the rotation groups in five dimensions, with different topologies: $SO(4, 1)$ refers to

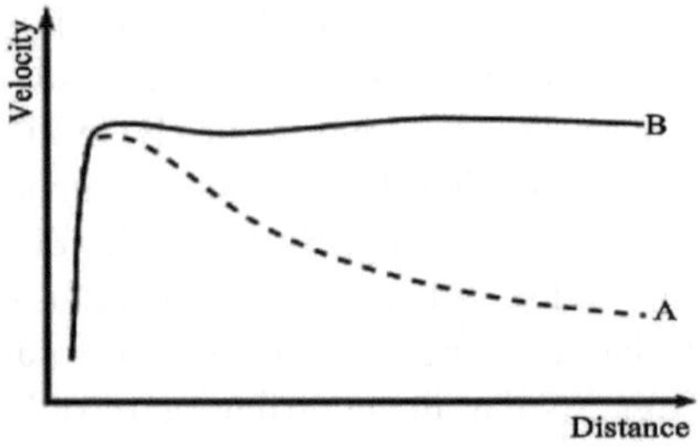

Fig. 2.6 A typical example of the star rotation curve in a spiral galaxy

the de Sitter metric signature $(++++-)$ and $SO(3,2)$ refers to anti-de Sitter metric signature $(+++--)$ (with two times). The Poincaré translations that respond for ordinary matter are lost, but they could be recovered in the flat Minkowski space-time by using a limiting process called group contractions [39].

One interesting interpretation of the cosmological constant is that it represents the accumulated vacuum energy denoted by $< \rho_{vac} >$, resulting from the presence of the quantum field remnants from the radiation era of the universe, so that the cosmological constant would explain the accelerated expansion of the universe. According to such interpretation the value of the cosmological constant was set to be proportional to the vacuum energy of the universe since its beginning [40]

$$\Lambda = -8\pi G < \rho_{vac} > .$$

In this case, Einstein's equations would then be written as

$$R_{\mu\nu} - \frac{1}{2}Rg_{\mu\nu} = 8\pi\mathbf{G}(T_{\mu\nu} + < \rho_{vac} > g_{\mu\nu}). \tag{2.18}$$

However, as it turned out, the theoretical estimates of Λ are much too small to cancel the huge amount of the observed dark energy. In other words, the difference between Λ and $- < \rho >_{vac}$ remains. Up the time of this writing we still do not know what dark matter and dark energy are.

2.3 Gravitational Waves

The existence of gravitational waves was predicted by Einstein since the first presentation of his gravitational theory in 1915, using the linear approximation of his equations. Assuming that the Minkowski metric is disturbed by a first order power expansion by some symmetric field $h_{\mu\nu}$,

$$g_{\mu\nu} = \eta_{\mu\nu} + h_{\mu\nu}, \quad h^2_{\mu\nu} << h_{\mu\nu}, \quad g^{\mu\nu} = \eta^{\mu\nu} - h^{\mu\nu}.$$

Calculating the Christoffel symbols, the Ricci tensor $R_{\mu\nu}$, and Ricci scalar R from the above metric, after some simplifications Einstein found that his gravitational equation for an arbitrary source reduces to

$$\Box^2 h_{\mu\nu} - \frac{1}{2}h\eta_{\mu\nu} = 8\pi\mathbf{G}T_{\mu\nu}.$$

Then, assuming the trace-free condition $h = g^{\mu\nu}h_{\mu\nu} = 0$, Einstein obtained a linear wave equation for $h_{\mu\nu}$ defined in a special coordinate system (the de Donder coordinates) such that $g^{\mu\nu}\Gamma^{\rho}_{\mu\nu} = 0$:

$$\Box^2 h_{\mu\nu} = 8\pi\mathbf{G}T_{\mu\nu}. \tag{2.19}$$

Since the Newtonian gravitational constant **G** has been assumed to be universal, then the hierarchy of the gravitational field also extends to the early stages of the evolution of the universe, meaning that the sources of gravitational waves described by Eqs. (2.19) are associated with Newtonian masses built in the source term $T_{\mu\nu}$. If, on the other hand, we consider gravitational waves originated from vacuum solutions, like, for example, a Schwarzschild black hole, then the radius of such black holes is given by $r_{Schw} = 2\mathbf{G}M/r$, where M denotes an integration constant which is necessarily identified with Newtonian mass of the Newtonian limit of the theory. Such black holes are generally described by as a massive black holes, where in General Relativity the presence of the Newtonian gravitational constant may be understood as meaning a large concentration of ordinary (atomic) matter.

The first experimental attempt to the local measurements of gravitational waves was made by Joseph Weber in 1960, using a mass-dipole antenna. Similar resonant mass detectors, including the use of vacuum enclosures and fine-tune thermal setups were constructed in different places and were tried in many other places, without success.

The first measurable evidence of gravitational waves was obtained by the astronomers Alan Hulse and Joseph Taylor in 1974. Since gravitational waves transport energy, Hulse and Taylor showed that in the binary pulsar 1913+16, there was an observable orbit decay, which corresponds exactly to the loss of energy as predicted by Einstein's gravitational wave emission [41].

A successful measure of the energy loss carried away by a gravitational wave emission was observed in 2015 by the LIGO Laboratories [42]. The LIGO laboratory was initially composed of two detectors located at Hanford and Livingston (respectively, in both Washington State and Louisiana State, both in the USA) at a distant of 3000 km apart. Each detector has four coplanar mass probes, two in each of the orthogonal axes, attached to mirrors like in the Fig. 2.7

Since each of these detectors defines two two-dimensional planes, then by themselves they cannot locally detect gravitational waves. That is, each of these detectors by themselves cannot detect all eigenvalues of the Riemann tensor. However, considering two such detectors separated by 3000 km, then it is possible to calculate the loss of energy carried by the gravitational waves from one location to another. The oscillations of the mass probes produce a periodic displacement of the pairs of masses. Using a Bayes statistical analysis, the measured differences are compared with some believable sources that presumably produced the observed periodic mass displacements. In 2015 the Hanford and Livingston instruments indicated a periodic displacement consistent with the hypothesis of a collision of two giant black holes' solution of Einstein's equations in General Relativity: The two instruments detected similar periodical waves within a fraction of seconds), with a loss of amplitude proportional to the energy loss along the 3000 km distance between the instruments. The result was compared with a traceless-transverse-plane-polarized simulation of gravitational wave produced by a collision of two giant massive black hole-like objects, with masses equivalent to 29 and 36 solar masses, respectively, estimated to have occurred at about 3 billion years in the past (Fig. 2.8).

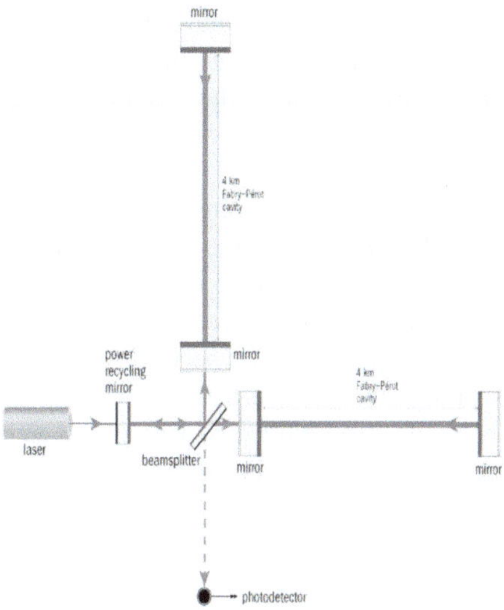

Fig. 2.7 The LIGO experiment

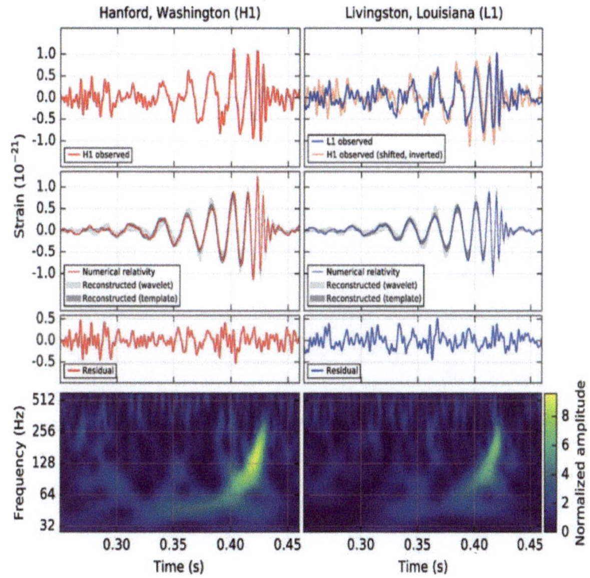

Fig. 2.8 The first LIGO results

Further results obtained by LIGO have motivated the construction of similar detectors in other places around the Earth, including the Virgo detector in Piza (Italy) and more recently the Kamioka detector in Gifu (Japan), all operating in the same basic principle. More recently, it has been noted that these detections were also (and unexpectedly) associated with other observed astronomical phenomena, in what became known as the "multi-messenger" phenomenology.

It is interesting to note that these instruments are capable of detecting a change in distance between its mirrors equivalent to 1/100th the width of a proton. Therefore at least in principle they have more than the required resolution, capable of measuring the wave-particle duality of the gravitational field at quantum domain. Consequently they may have also the resolution required to detect quantum gravity and black holes, provided they are properly set and tuned to do so with a high energy experiment.

Two additional points must also be commented: First, the linear gravitational field $h_{\mu\nu}$ maintains the same degrees of freedom of Einstein's gravitational field described by (2.19), so the observed wave must represent a spin-2 field. If so, then the emission and detection of gravitational waves in principle requires a quadrupole antenna [43, 44]. The second point is the fact that the linear wave equation (2.19) is a manifestly hyperbolic differential equation, so that the existence of waves must be confronted with the principle of general covariance. This second point is more complex and will be discussed in the next chapter in conjunction with quantum gravity.

Chapter 3
Quantum Gravity

Abstract The search for a theory of quantum gravity is almost as old as Einstein's General Relativity itself. Here we show that the hyperbolic characteristics of Einstein's equations cannot be neglected under the penalty that the Hamiltonian vanishes. In addition, we show that the existence of a nonvanishing Hamiltonian is a necessary and sufficient condition for the existence of a quantum version of the space-time geometry, consistently with the principle of general covariance.

Soon after Einstein's presentation of his theory in 1915, containing the example of the linear gravitational wave equation, Leon Rosenfeld concluded that *quantum gravity must exist*. However, as the history of the development of quantum gravity shows, more than a century later, including a wealth of models of quantum gravity, we still do not know what quantum gravity is, and here we to not intend to add another review. There are several good reviews on the specialized literature on this subject, as, for example, in [45–51], among many others.

However, the only concrete result explaining why it is difficult to obtain quantum gravity came in 1974, when G. 't Hofft and M. Veltman demonstrated that the gravitational field of General Relativity is non-renormalizable due to the presence of the Newtonian gravitational constant [52]. That result did not cause any immediate impact on the search for the canonical formulation of quantum gravity, possibly because its arguments are based on the renormalization procedure, which is an argument that is, alien from the canonical approach. Consequently, additional models of canonical quantum gravity proliferated anyway. However, it is not difficult to understand that the existence of many competing fundamental theories is worse than having none. The choice of one or another implies that gravitation in the Einstein sense would not be a fundamental interaction. In the next sections we will show some interesting examples.

M. D. Maia, E. M. Monte, *Geometry at the Quantum Scale*,
https://doi.org/10.1007/978-3-031-61197-1_3

3.1 Canonical Quantum Gravity

We start with the most direct way to Canonical Quantum Gravity, looking for the Schrödinger equation for the gravitational quantum wave function $\Psi(g_{\mu\nu}, \pi_{\mu\nu} \dots)$ and its associated momentum $\pi^{\mu\nu}$:

$$\frac{i\hbar}{c}\frac{\partial}{\partial t}\Psi(g_{\mu\nu}, \pi_{\mu\nu}, t \dots) = \hat{\mathscr{H}}\Psi(g_{\mu\nu}, \pi_{\mu\nu}, t \dots), \tag{3.1}$$

where $\hat{\mathscr{H}}$ denotes the "Hamiltonian operator" acting on the space of the functions $\Psi(g_{\mu\nu}, \pi_{\mu\nu} \dots)$ and t is the space-time time coordinate, all defined at the quantum scale.

Note that (3.1) is a (complex) wave equation that in principle (3.1) is consistent with the *invariant hyperbolic characteristic of Einstein's equations*, in which the derivative with respect to time is singled out. However, for the sake of the principle of general covariance, many researchers consider that they have right to write these same equations in any other non-hyperbolic reference frame. The problem is that when we do so, then the Hamiltonian vanishes, and consequently we also loose the whole Schrodinger quantum propagation. In other words, there is an *apparent incompatibility* between the principle of general covariance and the notion of quantum dynamics as associated with the Hamiltonian that needs to be resolved.

We say that the difficulty is apparent because from the point of view of functional analysis, Einstein's equations are globally hyperbolic. That is, these equations belong to the class of second-order system of equations in which the derivative with respect to one specific variable, say the time variable t, separates from the derivatives with respect to the remaining variables. Such property characterized as the determinant of the solutions [53–57]. It is precisely this hyperbolic characteristic that permitted Einstein to write the linear approximation of his gravitational wave equations as a wave equation. However in most regular texts on General Relativity, the hyperbolic property is frequently ignored, perhaps due to a non-justifiable suspicion that the hyperbolic property represents a compromise of the principle of general covariance of the theory. Actually, such compromise does not exist because the general covariance of the theory is already explicit in the Einstein-Hilbert action principle: The $\sqrt{-g}$ in that term is the same as the determinant $|J|$ of the Jacobian matrix of a generic coordinate transformation (for details on this see, e.g., [24])

$$\delta \int_{\Omega} \mathscr{L}_{EH} R\sqrt{-g}dv = \delta \int_{\Omega} R|J|dv = 0, \tag{3.2}$$

meaning that the principle of general covariance is already built in the Einstein-Hilbert principle. Consequently the hyperbolic characteristics of Einstein's equations do not impose any restriction on the principle of general covariance. This understanding was implicit in 1915 when Einstein derived the explicitly hyperbolic gravitational wave equation using the Donder coordinate gauge.

The same hyperbolic characteristics of Einstein's equations are frequently used without restriction in cosmology to study the expansion of the universe as a time parameterized foliation of tree-dimensional space-like hypersurfaces to explain the expansion of the universe, telling us that at every point of the space-time there is a three-dimensional hypersurface orthogonal to an integrable time-like vector field. In this sense the hyperbolic characteristic of Einstein' equations do not represent a conflict with the principle of general covariance of Einstein's equations. It simply says that the solutions of Einstein's equations can be "foliated" by three-dimensional space-like hypersurfaces: At every point of the space-time it is possible to choose reference frame where a time lapse experiment can be measured without ambiguity. The hyperbolic characteristic of Einstein's equations explains the *time problem* of Einstein's gravitation [58].

3.1.1 The Dirac Canonical Quantization

The hyperbolic characteristics of Einstein's equations have provoked also a profound disturbance in the understanding of quantum gravity. In 1959 Paul Dirac used the same hyperbolic invariant characteristics of the space-time to let an observer (a co-moving observer) move along an orthogonal time-like curved η. In such co-moving frame the space-time metric becomes simply

$$(g_{\mu\nu}) = \begin{pmatrix} g_{ij} & 0 \\ 0 & -1 \end{pmatrix}. \tag{3.3}$$

The Hamiltonian of the gravitational field is obtained from the Legendre transformation of the Einstein-Hilbert Lagrangian $\mathscr{L}_{EH} = R\sqrt{-g}$ as

$$\mathscr{H}_{Dirac} = \pi^{ij} g_{ij} - \mathscr{L}_{EH} = -\left[\,{}^3R - \frac{1}{{}^3g}\left(\frac{\pi^2}{2} - \pi^{ij}\pi_{ij}\right)\right]\sqrt{-{}^3g}, \tag{3.4}$$

where we have denoted by ${}^3R$ the Ricci scalar of the three-dimensional surface and by $\pi^{ij} = \frac{\partial \mathscr{L}_{EH}}{\partial g_{ij}}$ and $\pi = g^{ij}\pi_{ij}$ the momentum canonically conjugated to the space-like components of the metric. The Dirac Hamiltonian (3.4) is not zero, so that the co-moving observer sitting on the hyperbolic reference frame can in principle measure the quantum fluctuations of the space-time with an appropriate instrument [59].

The beauty of Dirac's canonical approach to quantum gravity resides in its simplicity and the correct application of Einstein's principle of general covariance. However, the hyperbolic characteristics of Einstein's equations is not a common sense. Some authors still believe that in spite of all that has been said, it represents a break of the principle of general covariance. Just like in the case of gravitational waves, the nonvanishing of the Dirac Hamiltonian is in full agreement with the hyperbolic characteristic of Einstein's equations. As such it represents a topological characteristic of Einstein's equations that does not break the principle of general covariance, which is built in the Einstein-Hilbert action principle, so that it does not preclude any observers. Actually it requires observers with clocks which are indispensable for any physical measurements of energy because all such measurements require an interval of time to be measured by some local clock. The real problem in the case of quantum gravity is that, contrasting with classical gravity, the observers (with or without clocks) do not belong to the quantum domain. Observers are classical concepts that live in the classical space-time. What changes in the case of quantum gravity and quantum space-times is that the observations are made though some distant purpose made measuring instrument (some kind of microscope or particle accelerators) that can probe the quantum space.

Actually, the necessity of measurement in unreachable regions by the observers of the space-time may explain why Einstein found it necessary to reinstate separately the principle of general covariance as an additional postulate of his theory: "All observers are equally good," with the meaning that all coordinate systems are admissible, *instead of excluding the hyperbolic co-moving observers*, used by Dirac in his quantum gravity and used in classical cosmology, as shown in Fig. 3.1.

The nontrivial Hamilton equations of Dirac's quantum gravity describe the quantum fluctuations of the space-time geometry with respect to time.

3.1.2 The ADM Theorem

In 1959 R. Arnowitt, S. Deser, and C. Misner (ADM for short) proposed a general canonical approach to quantum gravity, assuming that the principle of general covariance of General Relativity grants the right to make every reference frame suitable to define the quantum propagation of events. Thus, ADM suggested that the propagation of the quantum fluctuations of the space-time should not occur only in time, but it may also occur in any arbitrary direction in space-time, with a local time-like component $\eta^4 = N(x)$, where N is called the "lapse function," and space-like

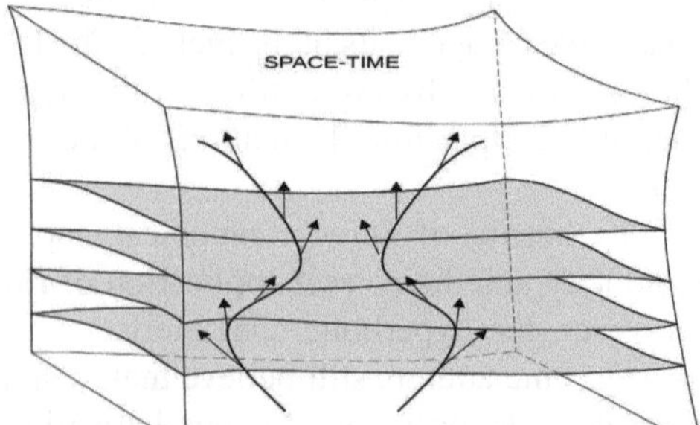

Fig. 3.1 "All observers are equally good" means also that hyperbolic reference frames are also admissible and not to be excluded

components $\eta^i = \frac{N^i}{N}$, where $N^i(x)$ is called the "shift vector" [68]. Consequently, in the ADM proposition the three-dimensional surfaces propagate along a general direction η with components $\{\eta^i, \eta^4\}$, in which the metric components assume the form

$$g^{\mu\nu} = \begin{pmatrix} g^{ij} + \frac{N^i N^j}{N^2} & \frac{N^i}{N^2} \\ \frac{N^j}{N^2} & -\frac{1}{N^2} \end{pmatrix} \tag{3.5}$$

and

$$g_{\mu\nu} = \begin{pmatrix} g_{ij} & N^i \\ \frac{N_j}{N_0} & -N^2 + g^{ij} N_i N_j \end{pmatrix}. \tag{3.6}$$

Using the same definitions of momentum canonically conjugated to the metric $\pi^{ij} = \frac{\partial \mathscr{L}}{\partial \dot{g}_{ij}}$ and its trace $\pi = g^{ij}\pi_{ij}$, they obtained the ADM Hamiltonian defined by

$$\mathscr{H}_{ADM} = \pi^{ij} g_{ij} - \mathscr{L}_{EH}. \tag{3.7}$$

After calculating separately the three Euler-Lagrange equations corresponding to the metric components g_{ij}, $g_{i4} = N_i$ and $g_{44} = N$, we obtain

$$\begin{cases} \frac{\partial \mathscr{L}_{ADM}}{\partial g_{ij}} - \frac{\partial}{x^\mu}\left(\frac{\partial \mathscr{L}_{ADM}}{\partial g_{ij,\mu}}\right) = 0, \\ \frac{\partial \mathscr{L}_{ADM}}{\partial N_i} - \frac{\partial}{x^\mu}\left(\frac{\partial \mathscr{L}_{ADM}}{\partial N_{i,\mu}}\right) = 0, \\ \frac{\partial \mathscr{L}_{ADM}}{\partial N} - \frac{\partial}{x^\mu}\left(\frac{\partial \mathscr{L}_{ADM}}{\partial N_{,\mu}}\right) = 0. \end{cases}$$

It follows that the first and third of these equations are consistent with the dynamics described by the Hamiltonian. On the other hand, the second equation involves the variation of the Einstein-Hilbert Lagrangian with respect to the shift vector N_i. As it happens, the application of the Euler-Lagrange equations with respect to N_i implies that the variation of ADM Hamiltonian necessarily vanishes

$$\mathscr{H}_{ADM} = -\left[{}^3\!R - \frac{1}{{}^3g}\left(\frac{\pi^2}{2} - \pi^{ij}\pi_{ij}\right)]\right]\sqrt{{}^3g} = 0. \tag{3.8}$$

Yet, the vanishing of the ADM Hamiltonian has led to a long and still open debate on the canonical approach to quantum gravity [60–62].

The persistent attempts to generalizing the Dirac Hamiltonian (3.4) with the presence of a nonzero shift vector are inconsistent with the Hamiltonian dynamics, because the invariant hyperbolic characteristic is an invariant characteristic of the solution of Einstein's equations.[1] Essentially, in quantum theory time and energy are naturally related by Heisenberg's relation, and this is exactly what is expressed in Dirac's canonical quantum gravity.

On the other hand the result expressed by (3.8) can be understood as the proof of a theorem, the *Dirac-ADM theorem*, stating that *the hyperbolic characteristic of Einstein's equations implies that the Hamiltonian of the gravitational field propagates only along the time direction.* While the general covariance refers to the structure of Einstein's equations, the hyperbolic characteristic is a characteristic associated with a solution of the same equations.

3.2 Loop Quantum Gravity

Another very popular approach to quantum gravity called Loop Quantum Gravity was initiated by Abhay Ashtekar and Lee Smolin in 1966. In brief, the process describes how quantum fluctuations of the space-time geometry may emerge from the quantum fluctuations of the $SU(2)$ gauge field [49, 50]. In 2010 the theory found a major difficulty associated with the emergence of the Mandelstam Inequalities, associated with the energy and momentum transfer in Loop Quantum Gravity [63]. A way out of such a problem was suggested through the use of Penrose's Spin-network combinatorial analysis [47].

From the point of view of the differential topology of the quantum space-time, Loop Quantum Gravity requires a geodesic triangulation of the space-time, as compared with the more common topological quadrangle of Einstein's space-time (to be detailed in Chap. 7) [61]. In order to understand the nature of the problem, it may be easier to start with a geodesic triangle in the space-time[2] formed by three geodesics passing through three points A, B, and C, connecting the points A to B, then from B to C, and finally from C back to A (Fig. 3.2).

Next, drag an arbitrary vector field $\mathbf{w}$ along the triangle A, B, C, obtaining a new vector $\mathbf{w}'$ at A which does not necessarily coincide with the original $\mathbf{w}$. Then the curvature of the manifold at point A is defined in differential topology by the difference between any such vectors

$$\mathscr{C}(\mathbf{u}, \mathbf{v})\mathbf{w} \overset{def}{=} \mathbf{w}'_A - \mathbf{w}_A = [D_{\mathbf{u}}, D_{\mathbf{v}}]\mathbf{w}.$$

[1] Back on the 1980s one of us (MDM) had a private conversation with Dirac at the ICTP, occasion in which we talked about his little book on gravitation and more specifically about his quantum gravity in special coordinates. Dirac commented briefly that the latter property is a direct consequence of the Riemannian structure of Einstein's equations for the space-times.

[2] A geodesic is a curve whose tangent vector $\mathbf{u}$ is transported parallelly along itself. In turn, the parallel transport of a vector field X along a geodesic is defined by the covariant derivative vector operator associated with an affine connection A with components $D_\rho X = \partial_\rho X + A_\rho X$.

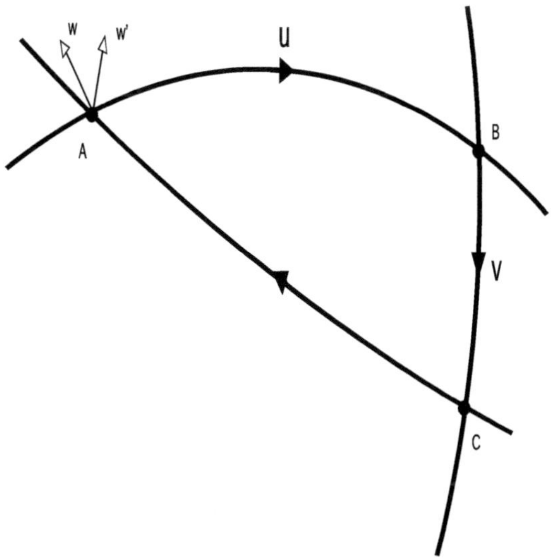

Fig. 3.2 Geodesic triangulation of a manifold

This is the definition curvature in Differential Topology. It coincides with the Riemannian curvature when it is expressed in terms of the Riemann-Christoffel symbols.

In the following we apply the above definition based on a single closed loop passing through an arbitrary point A, decomposed in three segments defined in the quantum space-time, as applied to the gauge symmetry SU(2). As we have defined in Chap. 1, that group acts on a three-dimensional manifold S_3 endowed with an affine connection A defined by Noether's theorem for the local $SU(2)$ gauge symmetry. Therefore, each component of the affine connection A_μ is represented by a 2×2 matrix, with components $A^i_{\mu j}$ defined in a basis of the three-dimensional internal manifold (Fig. 3.2).

Next consider the triangulation of the same S_3 manifold, but now composed by three segments of a *single closed loop* γ: The first segment has a finite length going from A to B, the second segment has an infinitesimal length $d\gamma$ connecting B to C, and finally the third segment denoted by $\gamma + d\gamma$ connects C to A, closing the loop as shown in the Fig. 3.3. As before, drag a triad of tangent and independent vectors (a reference frame) of S_3, and transport it parallelly along the segmented continuous loop again from point A, all the way back to A as shown in the next figure

The set of all above closed loops defines in fact a group, called the group of holonomy of the triad. As before, drag an arbitrary vector **w**, along the closed loops, to find that the result is not the same. The difference between the original vector at the point A, and the vector **w**′, transported along the closed loop, all the way back to the original point A, gives the differential topology expression of the curvature of the internal space S_3 at the point A.

$$\mathbf{w}' - \mathbf{w} = \mathscr{C}(\gamma, \gamma + d\gamma)\mathbf{w} = [D_\gamma, D_{\gamma+d\gamma}]\mathbf{w} = \mathscr{C}(\gamma, d\gamma)\mathbf{w}. \tag{3.9}$$

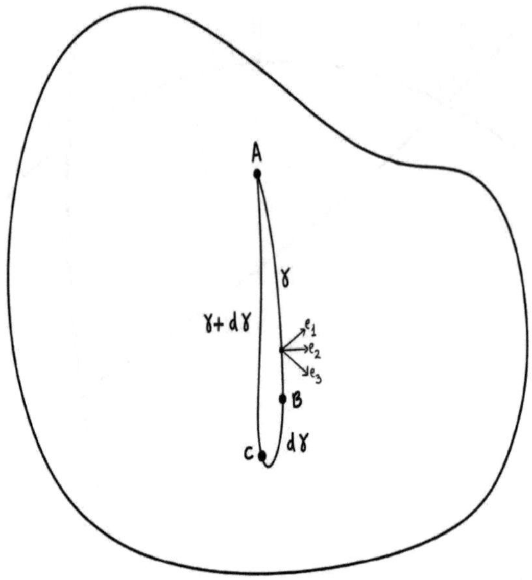

Fig. 3.3 The group of holonomy of a triad around a closed loop

Then, by taking the Taylor expansion of $\mathscr{C}(\gamma, \gamma + d\gamma)$ in powers of $d\gamma$, assuming that $|d\gamma|^2 << |d\gamma|$ gives

$$\mathscr{C}(\gamma, \gamma + d\gamma) = \mathscr{C}(\gamma, \gamma)d\gamma + \mathscr{C}(\gamma, d\gamma) = \frac{d\mathscr{C}(\gamma, d\gamma)}{d\gamma} + 0(d\gamma)^2.$$

Comparing with (3.9) after canceling the arbitrary vector **w**, we obtain a formal expression for the differential equation of the curvature operator along a closed loop as

$$\frac{d\mathscr{C}(\gamma, d\gamma)}{d\gamma} = \mathscr{C}(\gamma, d\gamma),$$

Which in principle may be formally integrated along the entire loop. Denoting the resulting curvature by $\mathscr{C}(\gamma)$, we obtain

$$\mathscr{C}(\gamma) = ce^{\int A_\mu dx^\mu}, \tag{3.10}$$

where c denotes an arbitrary integration constant.

Note that the expression (3.10) was derived for a single loop passing through a point. However, the geometry of the space-time may not be uniform around the chosen point, so that the above expression may need to be calculated for each loop passing through point A. Thus, for generality we should repeat the same for all possible closed loops passing through the same point A, defining a daisy-like picture

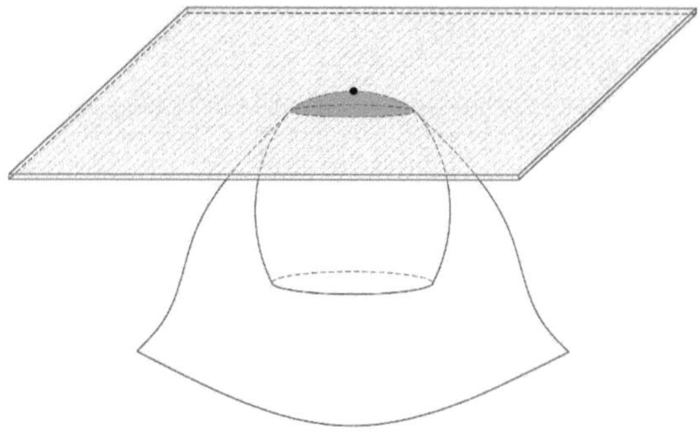

Fig. 3.4 Contact manifolds in loop quantum gravity

with infinite (and generally distinct) petals. On the other hand, for a sufficiently small and regular neighborhood, we may assume that for any given point we need to draw just one loop and then spin that loop around the same point.

The above construction is not trivial because, in general, the exponential in (3.10) involves the exponential of an algebraic function, like the exponential of a quaternion in the case of the $SU(2)$ or even like the exponential of anoctonion in the case of $SU(3)$. In both cases analytic exponential functions cannot be defined just like generalizations of analytic functions. Such a problem became evident at around 2010, when it was suggested that this difficulty would be solved with the use of Penrose's spin-network program.

Perhaps a more intuitive alternative to the spin network is to use the concept of higher order tangency between manifolds, shown by the shaded area in the Fig. 3.4 where the quantum fluctuations may be evaluated.

The theory of contact of manifolds is a well-established discipline that provides a more realistic description of what is happening at a single point at the quantum scale [64–67, 69, 70]. In the case of the quantum space-time, the quantum fluctuations induced by the gauge symmetry occur within the entire higher order tangent neighborhood around the tangent point. In this respect it should also be noted that in the higher order tangent manifolds the precise knowledge of the position of a tangency point is less relevant than the knowledge of what is going on in the neighborhood itself. For example, at the quantum scale the precision of the tangency point is not relevant due to the Heisenberg uncertainty relations.

In the example of the $SU(2)$ gauge field, instead of trying to cope with the behavior of the curvature at point A, we would be better off describing a wave-like quantum fluctuation of the contact region induced by the quantum fluctuations of the $SU(2)$ gauge field, as originally proposed by Ashtekar, simply because now we would be dealing with a well-established formal procedure to study quantum fluctuations of geometries.

Loop Quantum Gravity is an interesting proposal to quantum gravity that has attracted a substantial attention in the latter years. However, it also suggests that quantum gravity would be induced by gauge fields. If that is true, then Einstein's gravitation would not be a fundamental interaction.

Chapter 4
The Non-renormalizable General Relativity

Abstract One important result obtained in 1974 shows that Einstein's gravitational field of General Relativity is non-renormalizable, due to the presence of the Newtonian gravitational constant in Einstein's equations. Here we discuss the necessity of the renormalization of Einstein's theory of gravitation as an essential observational tool to define quantum gravity.

We start by contemplating the known scales of nature, showing how wide is the size of measurements of physical phenomena. The scales are given in measures of length (meters), temperature (Kelvin degrees), and energy (electron volts). The marked values in the scale of Fig. 4.1 are only approximate.

The first notable feature is the smallness of the human scale (the observers scale) as compared with the size of the universe and how large a human observer is as compared with the quantum scale. Also note that the so-called Planck scale, which is the present reference to compare physical theories, is located at a long way beyond the presently available observational energy, at the LHC,[1] something around $\sim$10 TeV. The implication is that even if the next decades we manage to build a larger and more powerful accelerator, we will not be able to reach that limit. This means that from the practical point of view of deciding which among the proposed theories of quantum gravity is the correct one, the Planck scale is useless.

As we have shown in the previous chapter, the Dirac canonical quantum theory of gravitation uses the Hyperbolic characteristic of Einstein's equations, which in principle allows for the construction of an explicit nonvanishing Hamiltonian operator, whose eigenvalue would indicate the true scale of relativistic gravitational energy. However, we have stressed the fact that most existing alternatives for canonical quantum gravity models exhibit an "apparent contradiction" between that hyperbolic characteristic with the principle of general covariance of Einstein's theory, simply because the time-like foliation evidences the existence of a privileged time-like coordinate.

[1] LHC stands for the Large Hadron Collider, the 32-kilometer particle accelerator located at the European Organization for Nuclear Research situated at the France-Switzerland border.

M. D. Maia, E. M. Monte, *Geometry at the Quantum Scale*,
https://doi.org/10.1007/978-3-031-61197-1_4

Fig. 4.1 The scales of physics

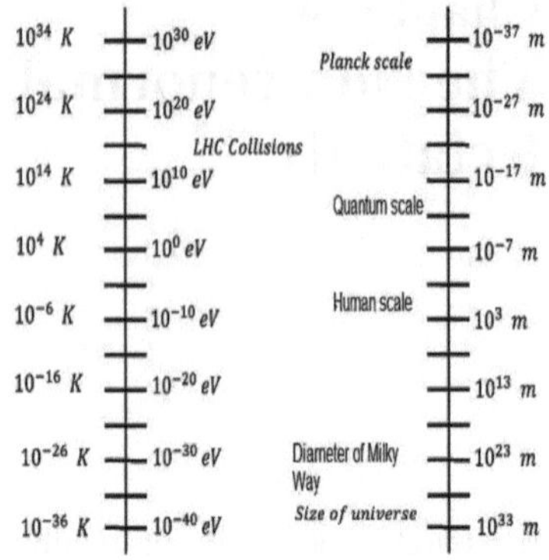

Indeed, that contradiction is apparent because the principle of the principle of general covariance sits at the higher level of the Einstein-Hilbert action, way before the hyperbolic characteristics are used to specify a particular solution of Einstein's equations, which always occur after a coordinate system is specified. What seems to be missing in the quantum gravity debate is the definition of the method of observations for quantum gravity. Why? Simply because the observer cannot access the quantum domain holding a metric scale in one hand and a clock in the other. The observers remain in their classical space-time while observing the quantum space-time at a distance. This is precisely the reason why we need to understand the necessity of renormalization of Einstein's gravitation.

Indeed, if the Dirac-ADM theorem solves the ambiguity of the canonical quantum gravity problem in Dirac's favor, then quantum gravity must be necessarily renormalizable. Said in other words, the renormalization of Einstein's gravitational theory is not an option and not an alternative process of quantization. Rather, the renormalization of a quantum field is a condition to ensure that the field remains finite when it is probed.

The renormalization is particularly relevant in nonlinear theories which are capable of self-interaction, because they may create or annihilate new solutions by themselves. Thus, they may need a process which guarantee that these newly created or annihilated solutions will remain finite. Only then we may check if a quantum theory can be observed at the quantum scale. If a quantum field is expressed in terms of a geometrical quantity like a metric or an affine connection, then the renormalization of that quantum field implies in what we call a renormalizable geometry.

4.1 Renormalization of Geometries

The renormalization of quantum field theory started in the mid-1940s, with the works of Hans Bethe [71], explaining the necessity to maintain finite the measurements at the quantum scale. However it took about 30 years until a proper

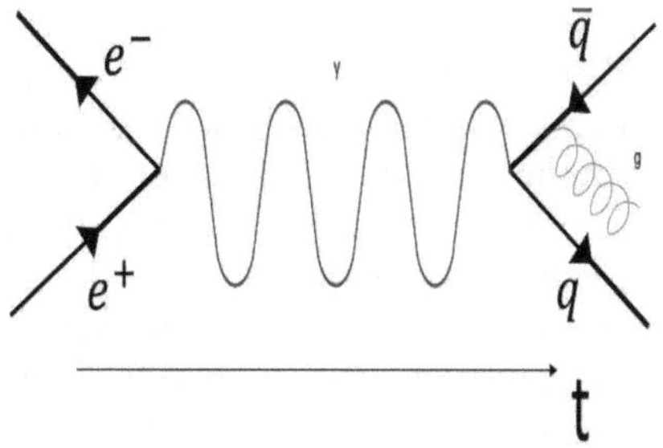

Fig. 4.2 A virtual electron-positron pair "annihilate," producing a photon γ, which subsequently "creates" a new pair of virtual particles. Note the time direction t indicated by the arrow

mathematical analysis of the renormalization process was defined, after the contribution of Kenneth Wilson [72–75], which eventually led us to the present understanding of the renormalization process, which is now a regular topic in graduate program on quantum field theory [76, 77].

As we have seen in Chap. 1, the gauge fields are associated with an affine geometry. In the study of the quantum gauge fields, their renormalizations imply that their corresponding affine geometries are necessarily renormalizable geometries. Likewise, the existence of quantum gravity necessarily implies that the geometry of the quantum space-time must also prove to be renormalizable, in the sense that the affine connection of the quantum space-time must also remain finite for all orders of perturbations of the metric. This is not really obvious, and at least in the case of General Relativity it has been proven to be impossible. Such understanding means that there is a serious conceptual problem with quantum gravity based on General Relativity.

The process of creation and annihilation of virtual particle pairs in quantum field theory was originally proposed in 1920 by Dirac, when he described how a pair of virtual particles composed of a spin-1/2 electron e^- and virtual spin-1/2 anti-electron (or positrons) interacts (or is *annihilated*) to produce spin-1 photon in the form of a radiation γ. Subsequently, that electromagnetic radiation decomposes into a new pair of virtual particles q^- and q^+ as shown in the Feynman diagram Fig. 4.2.

In principle the process of creation and annihilation of virtual particle pairs may continue indefinitely, thus producing what may be considered as a catastrophic (or explosive) condition which may or may not remain finite. Fortunately, such a process does not happen under natural circumstances, in which the output energy of such creation/annihilation of virtual pairs process remains finite and the output energy may be canceled by a power of the Planck quantum of energy $\varepsilon = \hbar\nu$. When this occurs, we say that the quantum process is renormalizable. All gauge forces have been proven to be renormalized [78].[2]

[2] The Feynman diagrams are part of an extensive diagrammatic language invented by Richard Feynman to explain in the early 1940s, before a complete theory, including the process of renormalization was fully understood, and before the emergence of a proper mathematical formulation of quantum field theory. Only later on, Feynman introduced his functional integral as a way to explain quantum field theory (see, e.g., [79]).

Much of the present renormalization procedure is due to Hans Bethe. Kenneth Wilson suggested a more intuitive approach, when he compared the renormalization process with the visualization of small scale objects. For example, like looking at a painting in a museum, under different light conditions of observation, so that the observer may visualize different details of the painting. Starting from the scale of the exhibition, in a well lit hall, the observer sees the classical view of the painting, as intended by the painter. However, as the observer gets closer to the painting, different details of the composition of the picture are observed: The overall meaning disappears, and eventually it loses whatever meaning it was supposed to represent. If the observer gets even closer, only some isolated blobs of ink on the canvas can be seen with the use of ordinary light (photons). If the observer insists in getting even closer, all that can be seen with the use of natural light (photons) as observational probes is the fiber structure of the canvas. It is possible to use a different light with shorter wavelength to a further look at fiber structure of the canvas. For an even more detailed observation, we need to take the picture to a high energy particle accelerator so as to observe the atomic structure of the canvas, thus getting a last glimpse of what we call ordinary matter composed of atoms. Beyond that, using probes with the highest available energy, after applying the energy conservation laws of quantum field theory, we may get an indirect picture of the nuclear structure of the atoms composing the painting. Such ultimate probing of matter is called a renormalization observational process.

Such simplified description of a complex high energy physics experiment to probe the nuclear structure of matter may sound somewhat naive. However, it is necessary and sufficient to understand the basic idea behind the high energy physics experiments, beyond the atomic structure.

Essentially, the renormalization procedure consists in perturbing the field by a sequence of values $\kappa_n \varepsilon^n \mathscr{A}^n\}$, for $n = 0, 1, 2, ...$ where κ stands for adjustable coupling constant of the self-interacting field, $\varepsilon = \hbar \nu$ stands for the Planck energy and $\mathscr{A}$ stands for the potential energy of the field in question, composing a power sequence like

$$\left\{\kappa_0, \kappa_1 \varepsilon \mathscr{A}, \kappa_2 \varepsilon^2 \mathscr{A}^2, ...\right\} \tag{4.1}$$

where κ_0 is usually assumed to be 1, corresponding to the lowest observational energy. The remaining coefficients κ_n are calculated from the analysis of convergence for each field separately. The relevant result obtained by Wilson is that such power sequence defines a group, called the renormalization group. It is somewhat intuitive to conclude that the above sequence formally defines an intuitive exponential given by

$$e^{\kappa \varepsilon A} \sim \sum_0^\infty \kappa_n \varepsilon_n A^n \sim \kappa_0 + \kappa_1 \varepsilon A + \kappa_2 \varepsilon^2 A^2 + ... \tag{4.2}$$

we say intuitive because the potential A in general is a matrix operator for which the above exponential does not necessarily represent an analytic function in the usual sense. This is a fine point that is often overlooked in regular textbooks on quantum field theory. Just to give an example, the exponential of a quaternion function which appear the electroweak interaction, in general is not a standard analytic function [80].

4.2 The Observational Limits of General Relativity

A very important result acknowledged by all was the proof by G. 't Hooft and M. Veltman in 1972, showing that all gauge field theories defined by the Yang-Mills equations are renormalizable, thus confirming the renormalization of the electroweak theory proposed by S. Glashow, A. Salam, and S. Weinberg [13, 14, 81], characterizing a unified theory of the electromagnetic and the weak nuclear forces with symmetry group $U(1) \times SU(2)$. The electroweak unification predicted the existence of two massive charged bosons $W^{\pm}$ and one neutral massive boson Z^0, whose existence was experimentally confirmed in 1983 at the CERN (today's LHC).

After their renormalization result in 1972, G. 't Hooft and M. Veltman made a similar analysis to the gravitational field of Einstein's theory of General Relativity in 1974, concluding that Einstein's gravitational field *cannot be renormalized* [52]. Ten years later, the same conclusion was reached using computers [82], leading to the same conclusion. More specifically, these results tell that the gravitational field of General Relativity is non-renormalizable. More to the point, these results indicate that the problem resides in the presence of the Newtonian gravitational constant **G** in Einstein's equations.

In order to understand the above conclusions and their implications, let us recall that in general the Lagrangian for an arbitrary physical quantity is defined in a region Ω with a boundary $\partial\Omega$ of the *space of configuration* of the system, in which the physical field Ψ and its first order derivative $\mathscr{L}(\Psi, \Psi_{,\mu})$ are well defined and are considered as *independent variables*. Then, the variational principle states that the action integral defined in Ω with respect to the total variation of Ψ is invariant

$$\delta \int_{\Omega} \mathscr{L}(\Psi, \Psi_{,\mu}) dv = 0...$$

As we know, the Euler-Lagrange equations, together with other fundamental characteristics of the field Ψ, such as the groups of symmetries of the action integral, the boundary conditions, and the definition of the observables, of the observers, and of the methods of observations, produce the correct Euler-Lagrange equations. The boundary conditions defined in a given boundary $\partial\Omega$ are necessary to prove the *existence and uniqueness* of the solutions. The problem lies precisely here. In the case of Einstein's General Relativity, the limiting boundary was defined after Planck imposed a further condition, namely $\mathbf{G} = \hbar = c = 1$, which defines what became known as Planck scale (shown in Fig. 4.1). The Planck scale is often referred to as the Planck regime or the Planck's geometric scale defined in Planck's theory of heat radiation [83].

Since **G** was adopted by Einstein from Newton's gravitational law without any modification, to his new theory of gravitation, then **G** carries its Newtonian characteristics to the Newtonian gravitation. However, since in Newton's theory the strength of the gravitational field is represented by a force, while in Einstein's theory, the strength of the gravitational field is defined by the curvature of the space-time. They represent different measurable quantities with different meanings, more specifically, if we write Newton's gravitational equation taking the units of mass, distance, and force. The gravitational field strength is given by (here the square brackets mean that only the unit values count)

$$[F] = [\mathbf{G}]\frac{[M]^2}{[L]^2} \quad \text{or equivalently,} \quad [G] = \frac{[F][L]^2}{[M]^2}, \tag{4.3}$$

which is very different from the Einstein notion of gravitation which is measured by the space-time curvature in General Relativity, in which each plane section of the space-time, is given by the ratio of a length measured in the tangent plane divided by the length measured in normal plane, producing a pure number.

The incompatibility of **G** with quantum gravity was demonstrated by 't Hooft-Veltman, showing that the problem resides in the squared mass term $[M]^2$ in the denominator of [**G**], telling us that the divergences may be canceled only up to the second order term in the renormalization sequence (4.1). In other words, the 't Hooft-Veltman conclusion tells that the gravitational field of General Relativity is non-renormalizable because of the presence of **G** in Einstein's equations. Therefore, the quantum gravitational field of General Relativity (with **G**) will not produce a stable quantum gravity, independently of the model, as long as the Newtonian constant is a component of the coupling constant with the source term (note that even for the vacuum solutions **G** appears as the integration constant). As we have already commented, the study of cosmology with the presence of **G** is always gauged by Newtonian matter, composed of atoms and their aggregates composing ordinary matter. Most of the above arguments have been absent in the debate on what model of canonical quantum gravity we should use. Indeed, not everyone working on General Relativity is convinced that we should remove **G** from Einstein's equations.

4.3 The Lifschitz Triple Point

In an effort to make the Planck regime something physical, in 1942 E. Lifschitz suggested an analogy with the concept of phase transition in condensed matter physics. In the same sense that it is possible to simultaneously observe ordinary matter (e.g., water) in three different states: as a solid (ice), as a vapor state

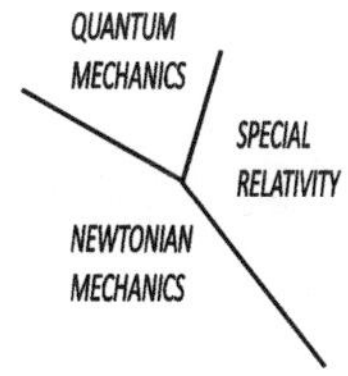

Fig. 4.3 The Lifschitz triple point

(steam), and as a liquid (liquid), Lifschitz suggested that it would also be possible to conceive "different states of physical theories" defined by their respective groups of symmetry. Thus, at least in principle we could have regions of space-time where Newton's gravitation combined with Special Relativity, Newton's gravitation combined with Quantum Mechanics, and Special Relativity combined with quantum Mechanic. In addition, the three regions share a common point known as the Lifschitz triple point, in which all three symmetries apply as shown in the Fig. 4.3 [84].

Obviously the Lifschitz triple point in Fig. 4.3 cannot be accessed by a classical observer because it would require a very high energy detector operating at an energy level of 10^{15} TeV and/or an instrument capable of measuring distances of the order of 10^{-33} cm and/or a clock to measure time intervals as short as 10^{-44} seconds. If such a level of resolution exists, it may be located somewhere and occurs in such a short interval of time that is very distant from our present reality.

4.4 Affinities and Their Connections

The existence of a geometry defined at the quantum scale is a direct consequence of the demonstration of Noether's theorem for local gauge fields. As we have seen in Chap. 1, the gauge potential ($A^b_{\mu a}$) was identified with *the affine connection of manifold defined at the quantum scale*. Such identification occurs through the use of the *adjoint representation* of the gauge symmetry, in which the representation space coincides with the space of the Lie algebra.[3] The picture below is the first page of the original handwritten edition of the Marco Polo Book of Travels, showing the Polo's ship departing from the Piazza San Marco in Venice, as seen from across the Grand Canal.

The lack of proportions was very common in most landscape paintings from the middle ages. As it can be seen in Fig. 4.4, there was no sense of proportion: The persons in front of the houses are shown almost as big or even bigger than the adjoining houses.

[3] There are two points to remember here: First, each representation is defined by the choice of a representation space, and second, a representation of a Lie group can always be derived from a representation of its Lie algebra.

Fig. 4.4 A picture drawn by Rustischello de Pisa ca 1200 gives a fine example of a space-time as seen from a distance, without the use of a properly defined covariant affine connection

The concept of affine geometry precedes that of Riemann geometry by about 150 years. It was originally introduced by Filippo Brunelleschi in Florence, Italy, at the beginning of the Renascence, and it was introduced as a way to measure objects in places where, for one or another reason, the observer cannot be located. Brunelleschi noted that even if the observer could not get closer to locally measure the sizes of the objects, a comparison between the observable with another known object with which there was some affinity could be established to define a sense of proportion. Thus Brunelleschi introduced the concept of *affine connection* at around the end of the middle ages, as a way to eliminate the discrepancy in sizes of the objects as seen from a distance, as long as all observers agree with that affinity. Such a concept of affine geometry has been frequently used in physics as, for example, in astronomical observations and in the internal spaces of gauge fields in quantum field theory. Therefore, the affine geometry seems to be also the appropriate instrument of observation for the space-time associated with quantum gravity, as seen from the observers that remain at the classical space-time, as long as it is consistent with the principle of general covariance of that theory.

This is precisely the same concept that has been used to define the geometry of gauge fields at the quantum scale in the analysis of Noether's theorem. If so, then Brunelleschi's affine connections are the key to understand Wilson's concept of renormalization. Indeed as we shall see, the use of an affine connection precedes, is more general and and it includes the Riemann concept of of metric geometry as a particular case. The very successful application of affine geometry to gauge theory strongly suggests that Brunelleschi's affine geometry is precisely what we need to define the space-time geometry at the quantum scale.

Before continuing it is important to understand that the Brunelleschi concept of affine connection is not unique in the sense that it does not specify a "standard affine connection." For example, in Fig. 4.4 we see that houses could be compared with people in the picture. They could also be compared with boats or even with swans. The concept of an affine connection is not unique. However once an affine connection has been chosen it must be applied to all observations. We shall see that there is a particular affine connection to be defined for the quantum space-time which is built in the Einstein-Hilbert principle.

Affine geometry precedes and is more general than metric geometry. Once an affine has been agreed upon, it must be kept and applied to all observers. For example, in cosmology it was found that three stars with the same redshift belong to the same three-dimensional space-like hypersurface. Therefore, by taking the sum of the internal angles of the triangle formed by the three stars with the same redshift, if that sum is respectively less, equal or greater than 180 degrees, then we may determine if these surfaces are hyperbolic, flat or parabolic as shown in Fig. 2.3.

The present measurements applied to the classical FRLW standard model of the universe described by a perfect fluid in Chap. 2 have been systematically found that universe is flat at least within the limits of the present observations [28]. However, as we know today, the FLRW model of universe cannot be a true representation of the physical universe because it is basically modeled as a classic gravitational theory, in the sense that it does not take into account the other fundamental interactions. For that purpose we would need something like a new concept of *quantum cosmology*, which would include all known fundamental interactions of physics, gravitation included, without hierarchies. Quantum cosmology is cosmology defined at the quantum scale, involving all known fundamental interactions, including the quantum gravitational field in which the natural measuring instrument is a general covariant affine connection, which is associated with the group of diffeomorphisms of the space-time.

The above concept stems from what has become known as Brunelleschi's experiment. The original experiment consisted of a small painting of the Florence baptistery with a small hole at the bottom, so that the painting and the real object can be seen from different positions by different observers while comparing the real object with the painted object (the baptistry) from any position. Interestingly, to improve the sense of reality, Brunelleschi burnished the top of the painting with silver, so that the real and the painted objects could be seen from the image reflected by the mirror, thus correcting the mirror (or chiral) symmetry as illustrated in the Fig. 4.5 [85].

Since the observers can move around and translate to different distances from the baptistery, these observers may compare the features of the observed baptistery

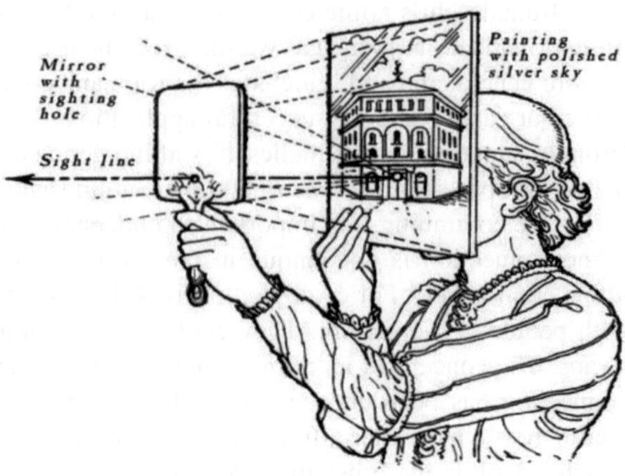

Fig. 4.5 The Brunelleschi experiment

with the affine objects or persons nearby. Note that in the original observed object in the figure, the Florence baptistery has an octagonal shape so that only a discrete rotation symmetry of 45° reproduces the same figure and unequivocally compares their relative sizes by affinity.

In the adaptation of the Brunelleschi experiment to the quantum scale observations, the internal quantum space is inhabited by the particle physics of the Standard Model, by the quantum manifolds of the gauge fields and by the quantum space-time, using the respective affine connections. In high energy physics, instead of photons, the observers use high energy beams of particles, and instead of looking at a picture, they compare the conserved quantities after and before the high energy experiments. The affine connection is defined by the respective group of symmetries of the observed object, namely the gauge symmetries for the Yang-Mills fields *and the group of diffeomorphisms for the quantum space-time.*

Chapter 5
The Renormalizable Einstein's Gravitation

Abstract The Einstein-Hilbert Action Integral for the gravitational field and Einstein's relativistic principles are derived from the renormalizable perturbations of the Minkowski metric by a massless, spin-2 quantum field, both defined at the quantum domain.

From the fact that *all gauge forces are renormalizable*, we infer that a differential geometry exists within the quantum scale. Such a conclusion stems from the Frobenius theorem in differential topology, telling us that the components of *Gauge fields strengths* $F_{\mu\nu}$ coincide with the components of the curvature tensor of the spaces on which the gauge groups of symmetries act defined within the quantum realm. As we see, the renormalization of the gauge fields is also geometrically significant to understand the nature of quantum field theory.

We will see that Einstein's gravitational field represented by the metric geometry of the space-time can also be understood within the quantum domain, provided the impediments for its renormalization are removed. However, as it is known but perhaps not too well known, the gravitational field of General Relativity is non-renormalizable, as an exclusive consequence of the presence of the Newtonian gravitational coupling constant **G** in Einstein's equations [52, 82].

We start with an interesting but somewhat obscure and neglected episode of the late development of Einstein's theory of gravitation: Back on the 1950s Einstein was getting concerned about the presence of singular solutions of his equation (before they became popularly known as black holes). Then, he asked one of his assistants, Robert Kraichnan, who had a background on quantum field theory, if he could explain the physical nature of these singularities. Would they represent be something fundamental? Asked Einstein! After some time (and it is not clear how long) Kraichnan came back to Einstein's office with a report along the following lines: Using Einstein's own definition of the observables of the gravitational field as given by the curvature tensor of the space-time, Kraichnan concluded that Einstein's gravitational field corresponds to a quantum field with five independent, nontrivial

M. D. Maia, E. M. Monte, *Geometry at the Quantum Scale*,
https://doi.org/10.1007/978-3-031-61197-1_5

degrees of freedom, or equivalently, using the spin-statistics theorem, Einstein's gravitation corresponds to a spin-2 quantum field. If so, then, Einstein's theory of gravitation could in principle be extended to the quantum scale of observations as a relativistic spin-2 quantum field, which had been already derived about 10 years earlier, by M. Fierz and W. Pauli [86, 87]. It is not clear from the historical records at Princeton how long that complex explanation lasted, how detailed it was, and what arguments were used by Kraichnan.[1] Apparently, Kraichnan's arguments came quite as a surprise to Einstein, whose original approach to gravitation as we know was essentially of classical nature, based on the Newtonian gravitation. It is also quite possible that the Kraichnan arguments were understood by Einstein as a theory of gravitation defined in the Minkowski space-time background, which could have been interpreted by Einstein as a theory of gravitation within the Minkowski space-time, thus jeopardizing Einstein's brainchild of a completely geometric theory of gravitation associated with the space-time curvature.

An additional complication of Kraichnan's argument is that in its most general form the Fierz-Pauli Lagrangian contains a mass term, which would in principle describe what is called today *massive gravity*. If so, then gravitation would be a short range field. The recovery of Einstein's gravitation would be obtained as a possible zero mass limit of the "massive gravitation." However, the presence of mass has led to a strong argument against the derivation of Einstein's equations, telling that the presence of mass necessarily implies in the presence of particles with infinite self-energy, which remains even after the zero mass limit [88].

To complete this interesting episode in the development of Einstein's gravitation, it should also be noted that Richard Feynman was also impressed by the spin-2 quantum field approach to gravitation. Indeed, along his course of lecture on gravitation Feynman tried to derive the gravitational equations, using his own functional integral during his course of lectures. However, he never finish that in due time. This is at least what we see in the highly recommended preface written by John Preskill and Kipp S. Thorn in the published *Lectures Notes on Gravitation*[2] [93]. For some unknown reason, Kraichnan delayed the publications on his spin-2 gravitation until 1995, after S. Gupta at the university of Manchester (UK) independently published his own findings on the same subject [94–96]. Kraichnan is best known for his relevant contributions to the problem of "turbulence in nonlinear fluid dynamics."

[1] We want to express our thanks to Prof. J. D. Barrow for his remark on Kraichnan's very brief approach to his conclusions published 5 years later.

[2] More recently, at a seminar Prof. Alexei Starobinski commented that a spin-2 approach to gravitation had already been proposed by Matvei Petrovich Bronstein back in 1930s [89–92].

5.1 General Covariant Quantum Gravity Without Ghosts

Presently, General Relativity remains as the acknowledged theory of gravitation, mainly because of its agreements with the classic observations made at the beginning of the twentieth century and more recently because the measurements of gravitational waves from the LIGO laboratories and other similar devices can be explained in terms of General Relativity as originating from the "Collision of two Massive Black Holes," which may have happened sometime in the past of the Standard Model of Cosmology. These observations were also associated with other astronomical observations, confirming that something gravitational in fact happened, even if these instruments do not have the capacity of measuring locally the complete set of five observables of the Riemann curvature tensor of the classical space-time. Instead, the interpretation of the LIGO experiment makes use of the Bayesian believable hypothesis test,[3] suggesting that the classical gravitational field of General Relativity is sufficient to explain other mentioned cosmological observations, such as the previously mentioned Cosmic Microwave Background Radiation and the more recent observations of the universe at the infrared frequencies, suggesting that the matter/energy composition of the universe is more complex than just the 4.9% of ordinary atomic matter. In other words, the fine resolution capabilities of the contemporary measurements of displacement of masses in the LIGO laboratories (something of the order of 1/10,000th the diameter of a proton) in principle can be adapted to quantum cosmology. If that is true, then these instruments can in principle can be upgraded to be capable of measuring all five independent eigenvalues of the Riemann curvature tensor at the quantum scale (something like a three-dimensional spherical mini-LIGO antenna), adapted to detect the graviton at the TEV scale, say at the presently planned new LHC.

Indeed, the previously mentioned 't Hofft-Veltman result of 1974 tells that Einstein's theory of gravitation can be made consistent with quantum mechanics provided **G** is replaced by a renormalizable coupling constant. The argument is very simple: Since **G** is defined by the Newtonian classical concept of mass (which is composed of atoms), then the quantization of Einstein's gravitational field can be obtained by replacing the gravitational coupling constant by another constant compatible with the renormalization of Einstein's gravitational field. In order to understand how this works, we need to derive again the Einstein-Hilbert action integral, this time starting from the massless spin-2 Fierz-Pauli quantum field, thus bypassing the classical Newtonian cosmological paradigm, at the same time that it incorporates the principle of general covariance within the quantum domain. In

[3] In 1783 Thomas Bayes introduced a new theory of statistics that includes the concept of relative probability $P(A|B)$ of an event A to happen under the hypothesis that it was produced by a believable event B. The Bayes theorem tells that $P(A|B)P(B) = P(B|A)P(A)$, where $P(A)$ denoted the standard statistic probability function and $P(B) \neq 0$ [97]. Such a theorem is used today as a main argument to test hypothetic simulations, using fast and large distributed computational facilities, and it belongs to the present usage of what has been called Artificial Intelligence.

the following we review the derivation of Einstein's gravitation from a massless spin-2 quantum gravity and show why it is consistent with the renormalization at the quantum scale. As we have explained such derivation was already known since 1950, but it requires some comments regarding some known objections, described below:

(a) *The Einstein-Hilbert Lagrangian derived from the Fierz-Pauli is not consistent with the principle of general covariance* [98].
The general covariance of Einstein's theory is set in the Einstein-Hilbert action in the form of the term $\sqrt{-g}$ which coincides with the Jacobian matrix of the group of diffeomorphisms of the space-time. Therefore the general covariance of Einstein's gravitation is explicitly stated before anything else is done. However, it is true that the principle of general Covariance was explicitly and independently stated by Einstein to define who are the admissible observers of his theory: Namely, all observers are admissible.
In the derivation of Einsteins equations from the massless Fierz-Pauli quantum field, the perturbations of the Minkowski reproduces the whole Einstein-Hilbert action with the same Jacobian matrix expressed in terms of $\sqrt{-g}$. Consequently the principle of general covariance is again explicit in the gravitational theory as derived from the Fierz-Pauli field.
The Fierz-Pauli field is a quantum field, so that the obtained gravitational equations must be associated with quantum sources of gravitation also defined at the quantum scale, without the Newtonian gravitational constant: At the quantum scale there is no ordinary or atomic matter available to compose a source of gravitation. On the other hand, once the constant **G** is removed, the resulting gravitational field becomes renormalized without hierarchies.
Since the observers are and always remain classical observers, the Levi-Civita affine connection cannot be used directly as an observational tool to measure objects at the quantum scale. Instead, a general covariant affine connection Γ_ρ must be defined, consistently with the observations at the quantum scale of events, besides being compatible with the renormalization procedures for the gravitational field.
(b) *Massive quantum gravity would contain ghost particles with infinite self-energies which remain even after taking the zero mass limit* [88].
Therefore the existence of such particles is alien to the experimental evidences of the Standard Model, and any attempts to recover Einstein's massless gravitation from the zero mass limit of a massive theory of gravitation would not work.[4]
Indeed, in its most general case the Fierz-Pauli Lagrangian contains a mass term, so that it may represent some kind of *massive gravity*. Massive gravity has been explored by many authors as in [99–102]. However, these masses refer to a quantum theory of fundamental particles of spin-2, whose mass spectrum,

[4] The authors express their thanks to the late professor Stanley Deser for a detailed debate on that point back in 1984.

at least for some of them, had already been experimentally determined in high energy physics experiments. The problem resides in the experimental evidence that masses of fundamental particles have been *experimentally shown to belong to a discrete set of numbers, separated by unpredictable intervals.* As such, the mass spectrum of the fundamental particles of the Standard Model forms an exquisitely discrete set of numbers, for which we cannot define a proper mass limit. Consequently, the Boulware ghosts would not appear, but we also do not get the zero mass graviton. Apparently, we may have a massive gravitation that would exert some kind of short-range gravitational force (like the moon tides on the surface of the Earth). On the other hand, Einstein's long-range gravitational field must be obtained independently from the massive spin-2 field.

(c) *Gerard 't Hooft and M. Veltman have proven that General Relativity is non-renormalizable due to the presence of the Newtonian gravitational constant* **G** [52, 82].

The renormalization of Einstein's gravitation is a real and difficult problem whose solution will be described along the remaining of this book.

5.2 Renormalizable Perturbations of the Minkowski Metric

We start with the derivation of Einstein's equations and of the Einstein-Hilbert principle and the diffeomorphism invariance, independently of any preliminary reference to Newton's gravitation and independent of any hierarchy between the fundamental interactions.

The Fierz-Pauli spin-2 symmetric field $h_{\mu\nu}$ written in the Minkowski space-time is defined by the Fierz-Pauli Lagrangian

$$\mathscr{L}_{Fierz-Pauli}=\frac{1}{4}[-h_{,\mu}h^{,\mu}+h_{\nu\rho,\mu}h^{\nu\rho,\mu}-2h_{\mu\nu}{}^{,\mu}h^{,\nu}+2h_{\nu\rho,\mu}h^{\mu\rho,\nu}-U], \tag{5.1}$$

where $h = \eta^{\mu\nu}h_{\mu\nu}$ denotes the trace of $h_{\mu\nu}$ and U is the potential energy of the field given by

$$U = \mu^2(h_{\mu\nu}h^{\mu\nu} - h^2), \tag{5.2}$$

where μ is an arbitrary constant.[5]

The observables of the Fierz-Pauli field were defined to be given by the components $h_{\mu\nu}$ themselves. Therefore there are six independent observables. In order to be a spin-2 field, one degree of freedom needs to be eliminated, which is obtained by imposing that the trace of $h_{\mu\nu}$ vanishes:

$$h = \eta^{\mu\nu}h_{\mu\nu} = 0. \tag{5.3}$$

[5] Since (5.1) is quadratic and homogeneous in $h_{\mu\nu}$ and its derivatives, then we may multiply $h_{\mu\nu}$ by any arbitrary constant, and if wished we may extract that factor from the Lagrangian $\mathscr{L}_{Fierz-Pauli}$.

The Euler-Lagrange equations for $h_{\mu\nu}$ derived from (5.1) are the wave equations:

$$(\Box^2 - \mu^2)h_{\mu\nu} = 0,$$

which is an extension of the Klein-Gordon equations for $h_{\mu\nu}$ with a "“mass term”" identified with μ.

As in any quantum field, the measurement of $h_{\mu\nu}$ must be done at its state of minimal potential energy defined by

$$\frac{\partial U}{\partial h_{\mu\nu}} = \mu^2(h^{\mu\nu} - h\eta^{\mu\nu}) = 0 \Rightarrow \begin{cases} \mu = 0, & \text{in a particular case, or} \\ (h^{\mu\nu} - h\eta^{\mu\nu}) = 0, \Rightarrow h = 0, \\ & \text{in general, by definition.} \end{cases} \tag{5.4}$$

Note that the above results represent only a measurement condition, which does not affect the Lagrangian itself (i.e., we cannot take the solutions of (5.4) back into the Lagrangian). Note also that the last case is a consequence of the definition of the field itself always applies regardless if $h_{\mu\nu}$ has or not a mass. Therefore, we infer that *in fact two different and independent spin-2 fields to be considered.*
One is a long-range massless spin-2 field $h_{\mu\nu}$, satisfying the wave equation

$$\Box^2 h_{\mu\nu} = 0. \tag{5.5}$$

The other field represents a short-range massive spin-2 field described by (5.2).

Several massive spin-2 particles have been detected in high energy particle experiments, evidencing the existence of a massive quantum spin-2 quantum particles, but none without mass (see, e.g., [103]). It is important to note also that mass associated with the parameter μ in (5.2) is not a mass in the same sense as defined in the Newtonian gravitation, simply because here at the quantum scale there is no participation whatsoever of the Newtonian concept of mass (i.e., ordinary atomic mass) in the above equations, simply because these equations are defined at the quantum scale. Furthermore, the invariance of the Fierz-Pauli equations under the Poincaré symmetry implies that *such mass must be an eigenvalue of the Casimir mass operator of the Poincaré group.* Note also that the coincidence of (5.5) with the linear gravitational wave equation strongly suggests that Einstein's gravitational field may be obtained by higher order perturbations of the Minkowski metric by the massless spin-2 field $h_{\mu\nu}$, also defined in the Minkowski space-time, as it was originally shown by Einstein.

However a zero mass limit does not exist in nature. This is so because the masses of the elementary particles do not compose a continuous set. In Chap. 2 we described the mass as a distribution of "elementary masses" composed of indivisible particles distributed on a given volume, producing a continuous mass density ρ that appears in Poisson's equation (2.4) for the Newtonian gravitational field. However this mass density refers to aggregates of ordinary mass composed of atoms. On the other hand, the masses that compose a fundamental theory are supposed to be the masses of the elementary particles in which by experimental evidence in high energy physics experiments they do not form a continuous set. Instead, the set of fundamental particle masses (called the mass spectrum) is very discrete in the sense that it is not even uniformly spaced. Therefore we cannot even consider a mass that is as close to another mass as we would wish, which means that we cannot talk about fundamental mass limits.

Actually, there is a formal procedure called *lattice gauge theory*, which is an idealized model in which a discrete distribution of mass is used to obtain solutions of the Yang-Mills equations (see, e.g., [104]). In such a proposal, the notion of zero mass limit is used in the context of a statistical arguments to attain a solution. Lattice gauge theory has been used to solve nonlinear differential equation, which has helped to solve the nonlinear Yang-Mills equations. However, as it can be seen in Fig. 1.3, the top line in each box shows the masses of the fundamental particle, and as we may clearly see they do not form a continuous set of variables. These masses are experimentally determined, and as such compose a discrete set of measured data as if they are a true gift of nature. As such, we cannot do anything about it.[6]

The subject of fundamental mass limit can be described by what we call the Boulware-Deser theorem, telling us that the conjectured existence of zero mass limit of the massive spin-2 field will necessarily imply in the emergence of particles with infinite energy states (the mentioned ghost particles). The only possible way to avoid the emergence of such ghosts in real space-times is that the Fierz-Pauli Lagrangian (5.1) in reality corresponds to *two separate and isolated physical situations*: The massive spin-2 field in the one hand and the massless spin-2 field in the other. These two physical situations must be solved separately, and the solution of the latter cannot be taken as the zero mass limit of the former.

5.3 The Renormalized Einstein's Relativity

The mentioned observational results in cosmology and high energy physics, together with the 't Hooft-Veltman result of 1974, strongly suggest that Einstein's equations derived from the Fierz-Pauli quantum field must be renormalizable, because it does not depend on the classical Newtonian physics. Rather, The Fierz-Pauli field is already defined at the quantum scale, although the observers remain classical.

In order to understand this conclusion, consider initially the fact the Fierz-Pauli Lagrangian (5.1) is *quadratic and homogeneous* in the Fierz-Pauli field components $h_{\mu\nu}$ and their derivatives. Therefore, $h_{\mu\nu}$ can be multiplied by an arbitrary renormalization constant, say $\kappa\varepsilon = \kappa\hbar\nu$, where κ is an arbitrary coupling

[6] The authors would like to express their thanks to professors Catherine Goldstein and Jim Ritter at the Sorbonne (Paris, France), for an enlightening conversation on the subject of limits of natural numbers.

constant and ε stands for the Planck energy. Then, for each value of such a product, calculate the perturbation of the Minkowski metric at the quantum scale, obtaining a sequence of perturbations of the Minkowski metric by the Fierz-Pauli field.

The resulting limit of such a sequence represents the metric $g_{\mu\nu}$ as a spin-2 quantum field satisfying Einstein's equations for an arbitrary source. The result defines a gravitational field within the quantum domain, a remarkable result initially obtained by Kraichnan, followed by Gupta and many others. More interesting is the fact that these results satisfy the same Einstein-Hilbert action, but now defined at the quantum domain

$$\int_{\Omega} R\sqrt{-g}dv = \frac{1}{\kappa}\int_{\Omega} \mathscr{L}_{Source}dv. \tag{5.6}$$

Consequently the result expressed by (5.6) defines an *already renormalizable quantum space-time geometry* at the quantum scale, which is completely independent of the Newtonian gravitational constant. Consequently the above result is also free of the hierarchy with respect to the gauge fields also defined at the same quantum scale.

5.3.1 TeV Scale Quantum Gravity

The classical space-time where we, the observers, live and play is the same classical space-time defined by Einstein, including the hierarchical limitations imposed by the presence of **G** in his gravitational equations. This means that the classical observational method defined by the metric for the classical space-time must be different in quantum gravity, simply because the observers cannot access the quantum domain, much less carrying a local metric measuring tool.

Concerning the Planck scale, we should keep in mind that similarly to the speed of light c and the Planck constant $\hbar$, **G** is not a pure number. The problem resides in the fact that the constant **G** was introduced in Newtonian gravitation, so that the above "physical dimensionality" holds only under the specific transformations of the generalized Galilean group, given by [6]

$$\begin{cases} x'^i = \sum_1^3 a^i_j x^j + c(t)^i + d^i \\ t' = at + b \\ \phi' = \phi - \sum_1^3 \frac{d^2 c(t)^i}{dt^2} x^i, \end{cases} \tag{5.7}$$

where a^i_j and d^i denote three-dimensional rotation and translation coefficients, in a three-dimensional space, a and b are the parameters of the "absolute time" transformations, ϕ stands for the Newtonian gravitational potential, and $c(t)^i$ is called a "time boost parameter." This is the group of symmetry of Newton's gravitational law, and consequently it is also the group of symmetry of the dimensionality [G] of **G**. Consequently, [**G**] is not necessarily invariant under the group of diffeomorphisms associated with Einstein's General Relativity.

In order to understand the implications of the above symmetry of the physical dimensionality, note that Newton's law can be read as saying that the ratio between the gravitational force F times a squared length r^2 divided by the squared mass m^2 is constant: $Fr^2/m^2 = \mathbf{G}$. Such a ratio is maintained constant only under the generalized Galilean group (5.7), where the constant value of **G** has been determined mostly by local experiments with torsion balances to measure the attraction force between two massive macroscopic bodies.[7] Therefore, *the assumed universality of* **G** *is an experimental result that is restricted to the region of space defined by the symmetry of Newton's law, known as the generalized Galilean group of transformations. Outside that region, like in the quantum domain, the use of* **G** *may lead us, and it indeed does, to wrong conclusions.* In the following we will detail how we may get around such difficulty, using the principle of general covariance without altering Einstein's Relativistic paradigm.

5.3.2 The Observables of Quantum Gravity

Contrasting with the Newtonian gravitational theory, the observables of the Einstein gravitational field are not characterized by a force. Rather, they are characterized by the local curvature of the space-time, a geometrical property defined by the Riemann tensor, invariant under the group of diffeomorphisms of the space-time.

In two dimensions any Riemannian manifold is transformable by a conformal scaling into a flat space, so that Einstein's gravitation is not definable in a two-dimensional space-time. In a three-dimensional space-time, we may have a solution of Einstein's equations but with some limitations. For example, there is no gravitational wave propagation in a three-dimensional space-time. A genuine gravitational theory in the context defined by Einstein requires a space-time with a minimum of four dimensions. In Chap. 2 we have seen that the simplest expression of curvature tensor in four dimensions occurs in the particular reference frame defined by its eigenvectors $X^{\mu\nu}$ which are solutions of (2.13). After eliminating the Minkowski flat ground state of gravitation, we end up with five nontrivial eigenvalues, which is often referred to as having five degrees of freedom (dof=5).

If less than five eigenvalues are measured, the observer will lack a complete picture of the gravitational field. At the quantum scale the number five is particularly relevant because the spin-statistics theorem of quantum mechanics translates five degrees of freedom into the quantum spin-2 characteristics. Therefore, the presence of Einstein's gravitation within the Standard Model of interactions will appear as

---

[7] Only more recently it has been determined, again locally, using the more precise Casimir effect [23].

a fundamental particle, the relativistic graviton, with zero mass and spin-2, to be recognized as such by all classical observers.

In the previous section we have seen how Einstein's equations are derivable from the "'renormalizable perturbations of the Minkowski metric,'" by a sum of powers of the spin-2 Fierz-Pauli field. Here we show that the resulting Einstein equations defined at the quantum scale remain invariant under the same group of diffeomorphism of the classical Einstein equations. In order to see how the diffeomorphism invariance emerges independently of the physical scale, consider an arbitrary n-dimensional Riemannian manifold will be defined, with metric tensor $g_{\alpha\beta}$, which transforms under a general coordinate transformation of the group of diffeomorphisms as

$$g'_{\mu\nu} = \frac{\partial x^\rho}{\partial x'^\mu}\frac{\partial x^\sigma}{\partial x'^\nu}g_{\rho\sigma}.$$

Taking the determinant of this expression, we obtain $g' = |J|^2 g$, where $|J| = det(\frac{\partial x^\rho}{\partial x'^\mu})$ denotes the *determinant of a generic transformation of the group of diffeomorphisms of the space-time*. Then, it follows that $\sqrt{-g'} = |J|\sqrt{-g}$. Using the initial coordinate system x^μ as the identity transformation for which $\sqrt{-g} = 1$, then we obtain for any other arbitrary set of coordinates that $\sqrt{-g} = |J|$. As we see, *the term* $\sqrt{-g}$ *that appears in the original Einstein-Hilbert action is identical to the determinant of the Jacobian matrix of a generic coordinate transformation* up to a trivial choice of the identity transformation [24]:

$$\int R\sqrt{-g}dv = \int R|J|dv. \tag{5.8}$$

In other words, the diffeomorphism invariance, or if you like the principle of general covariance of Einstein's gravitation, is already built in the Einstein-Hilbert variational principle. However, this fact does not exempt us to impose and announce the *principle of general covariance*. It is still necessary to define who are the eligible observers. Since the group of diffeomorphism includes any coordinate transformations, we conclude that all observers are equally good.

5.3.3 The Diffeomorphism Invariant Affine Connection

In order to express quantum gravity in the sense of Einstein's gravitation, the principle of general covariance must also hold at the quantum scale. Since the classical observer cannot just walk into the quantum domain with a metric instrument, then another observational method is required for quantum gravity. The observation of quantum gravity can be obtained from an analogy with the Brunelleschi experiment Fig. 4.5. In that experiment Brunelleschi was concerned precisely with the same problem of general covariance: how to observe a distant object in a landscape (like the Marco Polo Painting Fig. 4.4 as seen by different observers.

The Brunelleschi experiment Fig. 4.5 can be easily adapted to quantum gravity, by just replacing the octagonal shaped Florence baptistery by the quantum space-time and replacing the observers at the Florence Cathedral square by the classical space-time of the observers. In addition the observer must replace the discrete symmetry of the octagonal Florence Baptistery by the diffeomorphism invariant Eisenhart's affine connection as the observational instrument, so that all observers may agree on what they are looking at. As a tensor field, all observers agree in sizes and shapes of what they see. Note the difference with Einstein's observer using the Levi-Civita affine connection. However they all agree when they finally look at the Riemann curvature tensor[8] [106–109]. Consequently, observables and observers should be all happy with such arrangement, as long as they stay in local classical gravitation using the same metric geometry.

Since an affine connection is not unique (as an example we remind that in classical space-times the use of an alternative affine connection is quite common. For example, in cosmology we use the redshift of a star as an affine connection to evaluate the curvature of the space-like section of the space-time. On the other hand, in gauge field theory we use the gauge field potential itself as the affine connection, defined by the adjoint representation of the gauge group.). More specifically in the case of the quantum space-time the observer cannot use the local metric tensor as a measuring instrument, and we may define a general covariant affine connection. The lesson comes from Riemannian geometry, where we see that it is possible to define a vector-like affine connection in a curved space-time, associated with the group of diffeomorphisms of the curved space-time, such that at the end we get the same expression for the Riemann curvature, without explicitly using the Levi-Civita affine connection.

In the following we show how the Levi-Civita connection can be combined so as to produce a vector affine connection (we follow mostly the detail in [24]).

Consider a Riemannian space-time manifold with metric $g_{\mu\nu}$, and its invariant with respect to the group of diffeomorphisms of the space-time, expressed by the vanishing covariant derivative: $\nabla_\rho g_{\mu\nu} = 0$. From the definition of the Christoffel symbols $\Gamma_{\mu\rho\nu}$, this may be expressed as

$$\frac{\partial g_{\mu\nu}}{\partial x^\rho} = \Gamma_{\mu\rho\nu} + \Gamma_{\nu\rho\mu}. \tag{5.9}$$

Using the fact that the trace of the metric is constant $g^{\mu\nu} g_{\mu\nu} = 4$, it follows that

$$\frac{\partial}{\partial x^\tau}(g^{\mu\nu} g_{\rho\nu}) = g^{\mu\nu} \frac{\partial g_{\rho\nu}}{\partial x^\tau} + \frac{\partial g^{\mu\nu}}{\partial x^\tau} g_{\rho\nu} = 0. \tag{5.10}$$

[8] A brief historical note may be in order: The Levi-Civita connection was introduced by Tullio Levi-Civita in 1914, just before Einstein presented his gravitational theory in 1915, using the "Christoffel symbols" $\{\mu\nu, \rho\}$ and $\{^{\rho}_{\mu\nu}\}$, which were introduced by E. Christoffel in 1869, associated with the notion of parallel transport. In 1915 Einstein adopted the Levi-Civita connection to express the curvature tensor in General Relativity.

After multiplying this by $g^{\rho\tau}$, we obtain

$$\frac{\partial g^{\mu\tau}}{\partial x^\sigma} = -g^{\mu\nu} g^{\rho\tau} \frac{\partial g_{\rho\nu}}{\partial x^\sigma}.$$

Applying this result in the right-hand side of (5.9) and using again the definition of $\Gamma^\rho_{\mu\nu}$, it follows that

$$\frac{\partial g^{\mu\tau}}{\partial x^\sigma} = -\left(g^{\mu\nu}\Gamma^\tau_{\sigma\nu} + g^{\nu\tau}\Gamma^\mu_{\sigma\nu}\right). \tag{5.11}$$

From (5.10) it follows that

$$g^{\mu\nu}\frac{\partial g_{\mu\nu}}{\partial x^\sigma} = -g_{\mu\nu}\frac{\partial g^{\mu\nu}}{\partial x^\sigma}.$$

Applying these expressions in the derivative of the determinant g of $g_{\mu\nu}$, it follows that

$$\frac{\partial g}{\partial x^\sigma} = g g_{\mu\nu}\frac{\partial g^{\mu\nu}}{\partial x^\sigma} = -g g^{\mu\nu}\frac{\partial g_{\mu\nu}}{\partial x^\sigma}. \tag{5.12}$$

Using the derivative of a determinant,

$$\frac{\partial(-g)}{\partial x_\mu} = 2\sqrt{-g}\frac{\partial\sqrt{-g}}{\partial x_\mu},$$

from which we define

$$\Gamma_\rho \stackrel{def}{=} \frac{\partial(log\sqrt{-g})}{\partial x^\rho} \tag{5.13}$$

so that Γ_ρ are the components of a covariant vector, and hence a *covariant derivative operator may be defined by the representations of the group of diffeomorphisms of the space-time*, given by

$$D_\rho \mathbf{w} = \mathbf{w}_{,\rho} + \Gamma_\rho[\mathbf{w}], \tag{5.14}$$

where $\mathbf{w}$ denotes a scalar, vector, or tensor in space-time, and $\Gamma[\mathbf{w}]$ denotes the directional derivative of $\mathbf{w}$ with respect to Γ_ρ.

In order to conform this definition with the Levi-Civita affine connection, we assume that $D_\rho g_{\mu\nu} = 0$, so that

$$D_\rho g_{\mu\nu} = \nabla_\rho g_{\mu\nu} = \partial_\rho g_{\mu\nu} + \Gamma_\rho[g_{\mu\nu}] = 0. \tag{5.15}$$

Then it follows that the directional derivative of the metric with respect to the metric components is

$$\Gamma_\rho[g_{\mu\nu}] = -g_{\mu\nu,\rho}. \tag{5.16}$$

Therefore the entire Riemann geometry can be translated in terms of the vector affine connection Γ_ρ, using the covariant derivative D_ρ. The result gives the usual expression of the Riemann tensor in differential topology. Reciprocally, given the metric $g_{\mu\nu}$, we may obtain the diffeomorphism invariant affine connection Γ_ρ characterized by a special combination of the Christoffel symbols,

$$\Gamma_\rho = g^{\mu\sigma}\Gamma_{\mu\sigma\rho},$$

$$\Gamma_{\mu\nu\rho} = \frac{1}{2}(\Gamma_\nu[g_{\mu\rho}] + \Gamma_\mu[g_{\nu\rho}] - \Gamma_\rho[g_{\mu\nu}]). \tag{5.17}$$

Such a vector affine connection is commonly used in the coordinate-free formulations of Riemannian geometry, as, for example in [26].

Finally, note that similarly to the arguments used by E. Noether in gauge field theory, the covariant derivative D_ρ may also be in principle derived directly as in differential topology.

Chapter 6
Symmetry Mixings

Abstract The absence of the Newtonian gravitational constant in Einstein's equations not only makes Einstein's gravitational field renormalizable, but it also eliminates the hierarchy between Einstein's gravitational interaction at the quantum scale and the gauge field interactions. However, it also means that the Poincaré symmetry at the quantum scale must combine with the gauge groups of symmetry, bringing back an old, almost forgotten and unsolved problem: the symmetry mixing problem.

6.1 The No-Go Theorem

Some readers may still remember an old problem, telling us that it is not possible to combine the Poincaré symmetry of the Minkowski space-time with the gauge groups in any arbitrary way. That classic problem affects directly the allocation of the mass of the particles in the Standard Model of Interactions. The solution of such difficulty has been put aside for about half a century, while we wait for any experimental evidence of some possible generalizations for space-time symmetries based on the question: What is the most general group of symmetries that describe the correct mass distribution of the observed elementary particles? [110]. In view of the theoretical complexity involved in some presently proposed alternative symmetries, it seems reasonable to start from the beginning, with a brief review of the symmetry combination problem, including a comparison between the Poincaré and one supersymmetric example.

The original symmetry combination problem was concerned only with the combination of the Poincaré group with the gauge symmetries, independently from the gravitational field. However, in quantum gravity the problem becomes more acute for the following reason: Since, for the sake of renormalization, the Newtonian gravitational constant needs to be removed from Einstein's equations, then the hierarchy between gravitation and the gauge fields is also removed. Consequently, the symmetry mixing will occur at the energy level of the Standard Model. That means that the solution of the problem must be experimentally verified at the presently existing TeV scale of energy.

M. D. Maia, E. M. Monte, *Geometry at the Quantum Scale*,
https://doi.org/10.1007/978-3-031-61197-1_6

It should also be noted that the removal of the gravitational hierarchy in Einstein's theory has considerable implications to cosmology. Indeed, the presently existing cosmological models are essentially described by classical solutions of Einstein's equations; therefore, the elimination of **G** and the consequent existence of a renormalizable quantum gravity mean that the cosmological solutions will also be affected by symmetry mixing problem. In other words, if for any good reason the Newtonian gravitational **G** constant is eliminated from Einstein's equations, then the combined gravitational force with the gauge forces will have an equal, say, not only in high energy physics but also in cosmology, which ceases to be a purely classical subject to became also an area of high energy physics. And reciprocally, without **G**, high energy physics necessarily will also acquire a cosmological branch.

In order to understand the full implications of the breaking of the gravitational hierarchy, we need to specify what quantum geometry is. More specifically, what is the simplest geometry that we may associate with the Standard Model. Such geometry is necessarily associated with a higher dimensional manifold, involving the one-dimensional internal circumference S_1 acted upon by the $U(1)$ electromagnetic gauge symmetry, the two-dimensional internal spherical surface S_2 acted upon by the $SU(2)$ weak gauge symmetry, and finally the four-dimensional affine space S_4 acted upon by the $SU(3)$ gauge symmetry of the strong nuclear force [111–113]. To these we also need to add the four-dimensional quantum space-time solution of quantum gravity.

The simplest quantum geometry constructed with the above spaces involved within the Standard Model is composed by the direct sum of the four above-mentioned quantum spaces, leading to a seven-dimensional gauge space, plus the four-dimensional curved quantum space-time associated with the quantum gravitational field, so that we end up with an 11-dimensional direct sum manifold. The simplest total symmetry group is given by the Lie algebra given by the Cartesian product of all four involved Lie Algebras:

$$\underbrace{\mathscr{P}_4 \times U(1) \times SU(2) \times SU(3)}_{\text{The 11-dimensional Lie algebra of the total Lie symmetry group}} \quad . \tag{6.1}$$

This is the simplest symmetry that may be associated with Standard Model of Interactions, with the relevant feature that it is composed of the gauge and Poincaré symmetries experimentally confirmed by high energy experiments. Nonetheless, since the mentioned symmetry mixing problem applies to *arbitrary combinations* of the Poincaré symmetry with the gauge symmetries, then in principle it also applies to the product (6.1).

In more detail, the problem originated in 1965, with a theorem that bears the name of its author, L. O'Raifeartaigh, stating that *if we combine the Poincaré Lie algebra in any arbitrary way with any other symmetry group of the Standard Model, then all particles of that model having the same internal spin (belonging to the same spin multiplet) will also have the same mass* [114, 115]. More specifically, given two particles of the Standard Model, say a and b with the same internal spin, with respective masses m_a and m_b, then they will necessarily have the same (squared) mass:

$$m_a^2 = m_b^2. \tag{6.2}$$

However, the Wigner classification of the Unitary and Irreducible Representations (UIRs) of the Poincaré group shows that the observed masses and spins of the particles in the Standard Model are defined by the eigenvalues of the two Casimir operators of the Poincaré group [116, 117]: The second order Casimir operator of the Poincaré group defined by

$$C^2 = \eta^{\mu\nu} P_\mu P_\nu, \tag{6.3}$$

defines the mass of the particles, while the third order Casimir operator defines the spin of the same particles:

$$C^3 = \eta^{\mu\rho} \eta^{\nu\sigma} L_{\mu\nu}. \tag{6.4}$$

Now, in quantum mechanics, the averaged value of an arbitrary operator $\mathcal{O}$ between two states $|a>$ and $|b>$ belonging to a unitary irreducible representations of an internal symmetry group is given by

$$\frac{< a|\mathcal{O}|b >}{< a|b >}.$$

In particular, given $\mathcal{O}$ to be mass operator $C^2 = P^\mu P_\mu$ and taking E to be an arbitrary operator of the internal symmetry group, then the squared mass difference between the states $|a>$ and $|b>$ is given by

$$m_a^2 - m_b^2 = \frac{< a|[P^\mu P_\mu, E]|b >}{< a|b >}.$$

After changing the positions of indices a and b and using the nilpotency of the translation operators in the above commutator, it follows that the result is always zero:

$$m_a^2 - m_b^2 = \frac{< a|P^\mu (P_\mu E)|b > - < a|(E P^\mu) P_\mu|b >}{< a|b >} = 0. \tag{6.5}$$

Therefore, *independently of any specific way in which the Poincaré and the internal groups combine together, the (squared) mass difference will always be zero*. This result is in complete disagreement with the observed masses in the Standard Model shown in the table of Fig. 1.3. Note that the result (6.5) does not specify any particular internal symmetry, and it does not depend also on how the internal and the Poincaré groups combine. That is the reason why the suspicion falls into the Poincaré group structure.

Several tentative solutions of the problem have been presented, without indicating a definitive solution [118–122], leading to the conclusion that there is something that does not fit within the Standard Model: Either the Poincaré group cannot be used or its Lie algebra structure does not hold true at the level of high energy physics or yet that the whole Special Theory of Relativity in which the Poincaré symmetry was defined is wrong.

Back in 1967 a bold and daring solution of the symmetry mixing problem was proposed, assuming that the concept of point particles could be extended to one-dimensional objects called open and closed strings and that Wigner's theorem for classification of masses and spins of the elementary system could be extended to a higher dimensional Minkowski space-time.

However, when such a proposal was quantized, then it was found that for the closed strings the dimension of the resulting Poincaré invariant Minkowski space-time was compatible only for 26 dimensions. It was also found that it was possible to lower the dimensionality of the space-time to ten dimensions, if the 26-dimensional Poincaré symmetry would be replaced by the supersymmetric group, a symmetry that had been previously suggested by the *supergravity theory*, an earlier attempt to quantum gravity, that was later proven to be non-renormalizable [123].

The new supersymmetric theory was called *The Supersymmetric String Theory*, defined in a space-time with ten dimensions, with nine space-like dimensions plus one time dimension. There is a long list of publications on the subject including classics [124–127]. In principle such a geometric structure would solve the symmetry mixing problem, but it is also implied in the emergence of other unpredicted problems: Early in its first stages of development, it was found that there were two different types of strings: the open (type I) strings having quarks at its ends and the closed (type II) strings. Since the type I strings could violate the charge conservation, a new symmetry group was added to the theory, called the *anomaly cancellation symmetry*. Further calculations indicated the existence of two independent types of anomaly cancellation groups and consequently two new and independent types of string theories called, respectively, the "heterotic string model," in which the symmetry canceling group would be either $SO(32)$ or the exceptional Lie group $E_8 \times E_8$. Therefore, by 1990 there were five different string theories called type I open string, type II_A, and type II_B closed rotating strings and the heterotic string types $H_{SO(32)}$ and $H_{E_8 \times E_8}$.

To cope with five different fundamental theories is as bad or even worse than having none. Therefore by 1996, the solution for such a conceptual problem came in the form of a unification of the five string theories within an 11-dimensional space-time, including a selective compactification mechanism, required to explain how to recover the five ten-dimensional string theories through the use of a purpose built compactification mechanisms. In principle the 11-dimensional theory should be unique and final. However, M-theory involves some new complications of mathematical nature and a great deal of imagination to understand the physics behind it. Currently, the following model is very popular.

6.2 The ADS/CFT Model of M-Theory

The two de Sitter solutions of Einstein's equations in four dimensions mentioned in Chap. 2, with a cosmological constant $\pm \Lambda$, have, respectively, the de Sitter group $SO(4, 1)$ and anti-de Sitter group $SO(3, 2)$ as groups of symmetry. In particular, the anti-de Sitter group (ADS for short) has a Lie algebra which is consistent with the supersymmetric group in five dimensions. Finally, since these solutions do not admit translations, in principle there would not be any impediments from the mass splitting theorem related to the Poincaré group.

On the other hand, the conformal symmetry was once considered as a possible generalization of the Poincaré symmetry [132, 133]. However, such consideration was later discarded because the conformal symmetry violates the principle of cause and effect. In more detail, using the electromagnetic wave equations, the 15-parameter conformal symmetry holds true only if the advanced potential solution is taken together with the retarded potential solution. As it happens, the presence of the advanced potential represents a break of the causality principle. However, the conformal symmetry emerged again in the so-called ADS/CFT model of M-theory, when the four-dimensional space-time is regarded as a submanifold embedded in a five-dimensional anti-de Sitter space-time, which in turn is regarded as a submanifold of a six-dimensional flat space with a group symmetry $SO(4, 2)$. As it happens, the latter group is a 15-parameter Lie group which is also isomorphic to the 15-parameter conformal symmetry group of the four-dimensional Minkowski space-time [126–133]. Such somewhat complex symmetry correspondence gave rise to the so-called ADS/CFT model of M-theory.

In order to envision the ADS/CFT model, consider the use of spherical coordinates as in Fig. 2.1, where the de Sitter/anti-de Sitter solutions are written as ($+$ for the de Sitter and $-$ for the anti-de Sitter)

$$ds^2 = -dt^2 + \frac{4}{r \pm \Lambda^2} dr^2 + \frac{4}{r \pm \Lambda^2}(d\theta^2 + r^2 \sin\theta d\phi^2).$$

Calculating the components of the Riemann tensor, we obtain, respectively,

$$R_{\mu\nu\rho\sigma} = \pm\Lambda^2(g_{\mu\rho}\, g_{\nu\sigma} - g_{\mu\sigma}\, g_{\nu\rho}).$$

These solutions can be seen as four-dimensional hypersurfaces with constant curvature embedded in two different five-dimensional spaces with the respective symmetry groups $SO(4, 1)$ and $SO(3, 2)$, both of which can be calculated from the generic five-dimensional Lie algebra

$$[L_{AB}, L_{CD}] = \eta_{AC} L_{BD} + \eta_{BD} L_{AC} - \eta_{AD} L_{BD} - \eta_{BD} L_{AC},$$

where $\eta_{AB} = \text{diag}(+, +, +, -, +)$ for the de five-dimensional de Sitter space and $\eta_{AB} = \text{diag}(+, +, +, -, -)$ for the anti-de Sitter space, respectively.

Introducing the notation $P_\mu = \pm\Lambda L_{\mu 5}$, we may write the corresponding Lie algebras of both groups of isometries in a form that resemble the Poincaré algebra:[1]

$$[L_{\mu\nu}, L_{\rho\sigma}] = \eta_{\mu\rho}L_{\rho\sigma} + \eta_{\nu\sigma}L_{\mu\rho} - \eta_{\mu\sigma}L_{\nu\rho} - \eta_{\nu\rho}L_{\mu\sigma} \tag{6.6}$$

$$[L_{\mu\nu}, \ P_\rho\ ] = \pm\Lambda(\eta_{\nu\rho}P_\mu - \eta_{\mu\rho}P_\nu) \tag{6.7}$$

$$[\ P_\mu, \ P_\nu\ ] = \pm\Lambda^2\eta^{\rho\sigma}[L_{\mu\rho}, L_{\nu\sigma}]. \tag{6.8}$$

Using these expressions, the de Sitter and anti-de Sitter solutions of Einstein's equations can be regarded as examples of supersymmetric solutions. However, for that purpose it becomes necessary to introduce a "grading" on the Poincaré Lie algebra, by defining a new operator Q satisfying the conditions

$$[L_{\mu\nu}, Q] = -\gamma_{\mu\nu}Q, \ \text{ and } \ [P_{\mu\nu}, Q] = \alpha\frac{1}{2}\gamma_{\mu\nu}Q, \tag{6.9}$$

where α is an arbitrary constant and where we have denoted

$$\gamma_{\mu\nu} = \frac{1}{4}(\gamma_\mu\gamma_\nu - \gamma_\nu\gamma_\mu). \tag{6.10}$$

Where γ_μ represents the Dirac Matrices. Now the Lie algebras are nonassociative. Instead, they satisfy the Jacobi identity $[A, [B, C]] + [C, [AB]] + [B, [C, A]] = 0$. consequently, using (6.9), the Jacobi identity yields to a "graded Jacobi identity":

$$[A, [B, C]] + [C, [AB]] + \epsilon[B, [C, A]] = 0,$$

where $\epsilon = -1$ for fermions and $\epsilon = +1$ for bosons. With such adaptation the commutators involving the translation operators P_μ of the Poincaré group become

$$[P_\mu, [P_\nu, Q]] + [Q, [P_\mu, P_\nu]] + \epsilon[P_\nu, [Q, P_\mu]] = 0.$$

From which, it follows that $\gamma_{55} = [\alpha^2 + \Lambda\eta_{55}] = 0$, so that $\alpha^2 = -\Lambda^2\eta_{55}$. Taking either the Minkowski space-time or the de Sitter space-time, it follows that $\eta_{55} = +1$, and $\alpha = i\Lambda$ is an imaginary number, so that the supersymmetric algebra is not consistent with the Poincaré Lie algebra or even with the de Sitter Lie algebra. However, it is consistent with the anti-de Sitter Lie algebra.

Since the conformal group acting on a four-dimensional space-time is isomorphic to the anti-de Sitter group acting on the five-dimensional, AdS_5 space-time, it suggested the so called ADS/CFT conjecture, stating that the five-dimensional anti-

---

[1] Interestingly, note that when $\Lambda^2 << \Lambda$ we may recover the Poincaré Lie algebra, by a process known as the Inonu-Wigner group contraction [39].

de Sitter space-time may provide a an example of a 10-dimensional compactification of the 11-dimensional M-theory represented by the product

$$AdS_5 \times S_5,$$

where the sphere S^5 is added to cancel the string anomaly. However we still need to show that the above ten-dimensional product space is also supersymmetric. Figure 6.1 represents an attempt to visualize the ten-dimensional string model resulting from a compactification of the 11-dimensional ASDS/CFT solution of M-theory.

The question about the causality principle follows from the fact that both the conformal group $SO(4, 2)$ and the anti-de Sitter contain closed time-like curves. To remedy these causality breaking situations, the physical space-time is taken to be represented just a small section of the five-dimensional AdS_5 space (the small shaded area shown in the Fig. (6.1) below).

As we can see, the conceptual changes required by M-theory to explain nature are formidable to say the least, at least when compared with our present experimental approaches to the understanding of nature with our present notion of cause and effect as one fundamental concept of physics (Fig. 6.1).

In the next two chapters we will describe an alternative geometry for the Standard Model of Interactions, using the Higgs mechanism as one of its structural components, while keeping the Poincaré symmetry in an 11-dimensional quantum geometry.

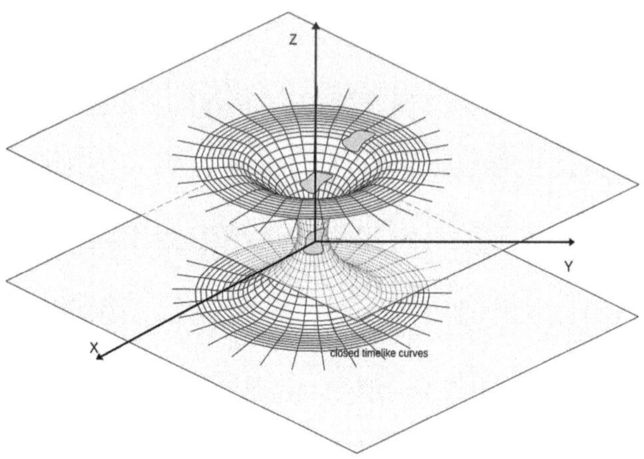

Fig. 6.1 Tentative view of an ADSCFT space-time. Note that the anti-de Sitter space-time has closed time-like geodesics like the circles

Chapter 7
The Four Geometries of the Standard Model

Abstract The Standard Model of Fundamental Interactions can only be considered complete after the inclusion of the gravitational interaction. In the present and next chapters we show how the quantum space-time geometry can be added to the three geometries associated with the gauge forces in an 11-dimensional quantum manifold, satisfying a classic system of nonlinear equations.

As commented in Chap. 2, using the de Donder coordinate gauge, Einstein demonstrated for the first time in 1915 that his gravitational equations have a hyperbolic nature (as compared with parabolic or elliptic characteristics). However, it took a century to convince most researchers on the importance of the hyperbolic characteristics of Einstein's equations after the direct detection of gravitational waves at the LIGO laboratories. The same hyperbolic characteristic remains true at the quantum scale, either from Dirac's canonical quantization of the gravitational field or from the inclusion of the gravitational field at the quantum scale using the 't Hooft-Veltman perturbative analysis of quantum gravity, associated with the renormalization of the Minkowski metric by a massless Fierz-Pauli quantum field.

From the physical point of view, the only difference between the classical and the quantum space-times lies in the method of observations. They must be different simply because the observers cannot have direct local access to the quantum domain. Similarly to what is done in gauge field theory, the observations of the quantum space-time must be made at a distance, using an appropriate affine connection, consistently with the principle of general covariance defined by the group of diffeomorphisms of the space-time.

As we know, the standard Levi-Civita affine connection used in the classical Riemannian geometry does not work because they are not tensors. If we insist in using the Levi-Civita affine connection at the quantum scale, then different observers using different coordinates will never agree on what they are looking at. As we have already discussed in Chap. 5, the solution for this problem is to use a combination of Christoffel symbols such that the result generates a vector affine connection (5.13) in the space-time. Note that such a general covariant vector affine connection is consistent with, and in principle it may be used with advantage

M. D. Maia, E. M. Monte, *Geometry at the Quantum Scale*,
https://doi.org/10.1007/978-3-031-61197-1_7

in classical General Relativity, although such use may be dispensed with, simply because in the classical space-time the metric is the observational tool.

Therefore, by taking into considerations all known quantum fields within the Standard Model plus gravitation, the simplest geometry that can be associated with that model is given by the direct sum of the four spaces where each of the four groups of *symmetry act*. The result is an 11-dimensional manifold defined at the quantum scale, with four different affine geometries, one for each of the four known fundamental forces.

In the next four sections we describe the geometry of the four fundamental interactions of the Standard Model and their associated algebras. The reader may want to compare the observations of the four quantum geometries with Brunelleschi's experiment described in (4.5), where instead of the Florence Baptistery, the observers are looking separately to each geometry associated with each quantum submanifold embedded into the total 11-dimensional quantum space. For each fundamental interaction we identify a geometry and a Lie algebra written in their respective adjoint representation as described in Chap. 1.

7.1 The Geometry and Algebra of Quantum Electrodynamics

The simplest quantum gauge field is Quantum Electrodynamics (QED). As we have seen in Chap. 1, it is defined by classic Maxwell's equations. However, its quantum version called quantum electrodynamics is defined at the quantum scale by the gauge symmetry group $U(1)$ with only one parameter which is a function of the space-time coordinates. Its associated Lie algebra is also one-dimensional, acting as a group of translations of a one-dimensional space, a line, denoted by S_1. This Lie algebra is composed of 1×1 unitary matrices, associated with the components of the quantum electromagnetic potential A_μ, defined in the quantum domain. You may want to compare these components with the components of the electromagnetic potentials in Noether's theorem described in Chap. 1 applied to the case of the symmetry group $U(1)$.

In order to visualize the representation space S_1, recall that the group $U(1)$ is isomorphic to the group of rotations $SO(2)$ acting on a one-dimensional circumference drawn in a two-dimensional representation space (a plane) of the group. Then the points of the group $U(1)$ are represented by the projections of the points x from the north pole (x) of the circumference by a point (y) in the horizontal line. This line is the representation space of the Lie algebra of the $U(1)$ symmetry, as shown in the Fig. 7.1:

Note that since the observer may parameterize each point of that circumference with coordinates (x, y), then the observer may also define a complex number $z = x1 + yi$ with norm $|z|^2 = z\bar{z} = x^2 + y^2$. In particular taking the North pole of the circumference shown in Fig. (7.1) to represent the identity element of the group of

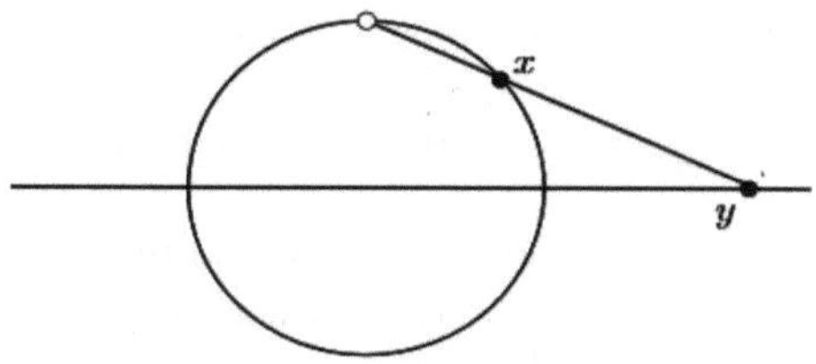

Fig. 7.1 The stereographic projection of a circumference into a line

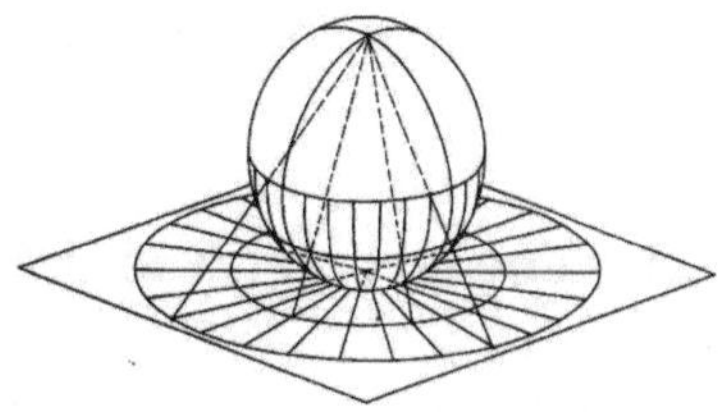

Fig. 7.2 The stereographic projection of sphere into a plane (from the North Pole) maps all so that all points of that sphere into a two-dimensional plane

rotations $SO(2)$, then the horizontal tangent line passing through N coincides with the space of the Lie algebra of the $SO2$ group. Equivalently, we may represent the same group by the group of automorphisms of the complex algebra.[1]

7.2 The Geometry and Algebra of the Weak Nuclear Force

In the case of the weak nuclear force, the symmetry group $SU(2)$ is isomorphic to the group of rotations $SO(3)$ acting on a two-dimensional spherical surfaces S_2, defined in the three-dimensional quantum space. However, differently from the electromagnetic example, the weak nuclear force has no classical correspondence. Therefore its representation space is defined within the quantum domain only. In order to define one such representation, we may generalize the previous case of rotations in a circumference to rotations on a spherical two-dimensional surface S_2 as shown above in the Fig. 7.2.

Similarly to the electromagnetic example there is also an associated algebra, the quaternion algebra that is a generalization of the complex algebra such that the points of the sphere can be represented by a quaternion described as

$$\mathbf{q} = t\mathbf{1} + x\mathbf{i} + y\mathbf{j} + z\mathbf{k}$$

[1] Since in this example the plane can be taken as a plane of the classical Minkowski space-time, then the representation of the gauge group by a rotation in the circumference seems to be trivial. However, remember that we are describing quantum electrodynamics, defined in the quantum space-time and not in the classical space-time.

with quaternion conjugate given by

$$\bar{\mathbf{q}} = t\mathbf{1} + x\bar{\mathbf{i}} + y\bar{\mathbf{j}} + z\bar{\mathbf{k}} = t\mathbf{1} - x\mathbf{i} - y\mathbf{j} - z\mathbf{k}$$

whose norm is defined by $|\mathbf{q}|^2 = \mathbf{q}\bar{\mathbf{q}} = t^2 + x^2 + y^2 + z^2$.

Therefore we may represent the affine connection of the weak nuclear force as a quaternion with components A_μ. Such representation is usually made by the set of Pauli matrices, which belong to a representation of the quaternion algebra defined at the quantum space-time by 2×2 matrices (see, e.g., [135]).

7.3 The Geometry and Algebra of the Strong Nuclear Force

Unlike the two previous cases the unitary symmetry $SU(3)$ is not isomorphic to any rotation group. Rather, it belongs to the class of Lie groups known as the **exceptional Lie groups**. There are five such exceptional Lie algebras, with increasing degrees of complexity denoted, respectively, by G_2, F_4, E_6, E_7, and F_8 [136, 137]. The Lie algebra of $SU(3)$ is isomorphic to the smallest exceptional Lie algebra G_2, related to the group of automorphisms of the algebra of Octonions $\mathbb{O}$ which is the largest known division algebra (the others are the reals, complex, and quaternions).

The octonion algebra is described by seven independent vectors plus the unit element of the algebra, usually written as

$$\mathbf{o} = t\mathbf{1} + x\,\mathbf{i} + y\mathbf{j} + z\mathbf{k} + l\,\mathbf{li} + m\,\mathbf{lj} + n\,\mathbf{lk} + o\,\mathbf{ijk}$$

satisfying also a well-known multiplication table, which is usually described by the Fano diagram shown below (Fig. 7.3): [138]

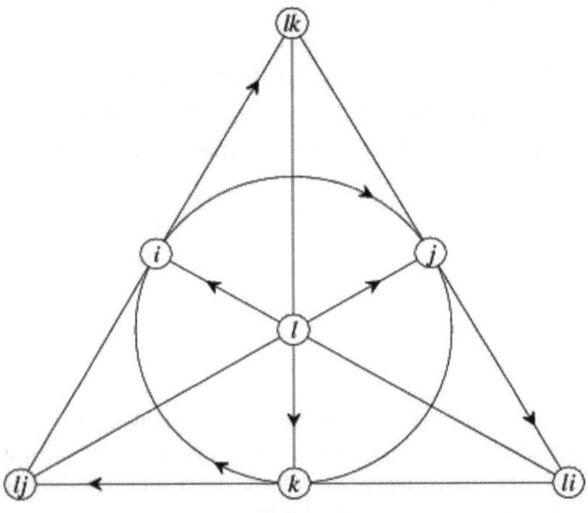

Fig. 7.3 The geometry of the SU(3) is described by four tangent spaces, composing a tetrahedron or a triangular pyramid

The Fano diagram is graphical a way to express the exceptional Lie algebra of the $SU(3)$. Following the two previous examples of gauge symmetry, the $SU(3)$ exceptional Lie algebra may be represented by a pair of quaternions in such way that it composes an octonion. Thus for example, in the octonion **o** shown above diagram the octonion can be written as a pair of quaternion as shown above. However, the choice of the **1** base can in principle be made with the base elements **i**, or **j** or **k**, so that the Fig. 7.3 can also be seen as a particular face of the mentioned tetrahedron, which encloses the exceptional geometry of the exceptional Lie algebra $SU(3)$ [139–141, 188].

7.4 The Geometry and Algebra of Quantum Gravity

To complete the set of four geometries that are defined at the quantum scale, we need to define the geometry of the quantum space-time. As in the previous examples, this geometry is determined by the classical observational properties, which in this case is the same group of diffeomorphisms of the classical space-time. The classical group of diffeomorphisms is an infinite Lie group whose infinite parameters are functions of the space-time coordinates. Consequently, the corresponding Lie algebra is also an infinite Lie algebra. As it was also seen in Chap. 5, the Einstein-Hilbert action principle includes the term $\sqrt{-g}$, which happens to be the same as the determinant of the Jacobian matrix of the group of diffeomorphisms of the space-time denoted by $|J|$. This determinant is independent of the representation of the group of diffeomorphisms of the space-time solution of Einstein's equations. Consequently, it is natural to take the representation of the group of diffeomorphisms of the space-time defined in the space of its Lie algebra, which means to use the adjoint representation of the group of diffeomorphisms of the quantum space-time, just like we did with the other three fundamental representations (if you remember, that was the choice made by Noether in her theorem to obtain the conserved quantities associated with the gauge fields.) With this choice, the general covariant affine connection Γ_ρ may be used by all classical observers to measure the four-dimensional quantum gravitational field. *The choice of the adjoint representation for all fundamental interactions* represents a complete agreement with Einstein's principle of general covariance which is therefore implemented in the quantum gravitational field, besides being a conceptual improvement over the adoption of the traditional Levi-Civita affine connection used in classical General Relativity.

Indeed, although it was first thought of in a different context, Brunelleschi's experiment may provide us with an intuitive idea of a general covariant affine connection: Different observers located in different positions of our classical space-time may agree on the observed curvature (or the quantum gravitational force) of the quantum space-time (which in this comparative example replaces the Florence baptistery in the original experiment as shown in Fig. 4.5, as seen by a classical observer, probing the quantum space-time.

However, in order to obtain the Riemann curvature tensor from the affine connection Γ_ρ, we need to rewrite the Riemann curvature in terms of Γ_ρ. This expression is given in differential topology by the Frobenius theorem, establishing the conditions for the integrability of the Riemann curvature [26, 55]. The only difference is that the local space-time metric is no longer the observational instrument, which was replaced by the affine connection $\nabla_\rho g_{\mu\nu} = 0$, and the corresponding derivative operator

$$D_{\mathbf{u}} = \partial_{\mathbf{u}} + \Gamma_{\mathbf{u}}, \tag{7.1}$$

where **u** denotes an arbitrary tangent vector. The condition imposed on (7.1), so that it agrees with the classical Riemann definition of covariant derivative associated with a metric, is

$$D_\rho g_{\mu\nu} = D_\rho < e_\mu, e_\nu >= 0.$$

Therefore, we may apply (7.1) to the quantum space-time, defining a geodesic in the quantum space-time as a curve whose tangent vector keeps always parallel to itself[2]

$$D_{\alpha'}\alpha' = 0.$$

In order to maintain the concept of *curvature as the observable of the gravitational field* within the quantum scale, we apply the Frobenius theorem on differential topology, stated as follows: Consider the four-dimensional quantum space-time endowed with the metric-affine connection Γ, and on it draw a geodesic parallelogram with vertices at the points $A,\ B,\ C,\ D$, defined by parallel geodesics as shown in the figure below:

Following the above picture, a geodesic α with tangent vector α' connects A to B and another geodesic β from B to C with tangent vector β'. To complete the figure draw a third geodesic from C to D very close to α with tangent vector $-\,alpha'$ and finally another geodesic from D back to A very close to β, with tangent vector $-\beta'$. In order to close this geodesic parallelogram, we impose the closing condition telling that the four lines compose a closed parallelogram (Fig. 7.4)

$$D_{\alpha'} D_{\beta'} = D_{\beta'} D_{\alpha'}.$$

Next we drag an arbitrary vector field **w** along the quadrangle from A and back to the point A, just to find that it does not necessarily coincide with the original w. The difference between these vectors defines the Riemann curvature of the manifold associated with the affine connection Γ, given by the difference

$$\mathbf{w}' - \mathbf{w} = \mathscr{C}(\mathbf{u}, \mathbf{v})\mathbf{w} = [D_{\mathbf{u}}, D_{\mathbf{v}}]\mathbf{w}. \tag{7.2}$$

[2] This is essentially the same definition of geodesic in classic Riemannian geometry, with the difference that it keeps the topological characteristic associated with the notion of parallel transport, which lends the name of differential topology of differentiable manifolds.

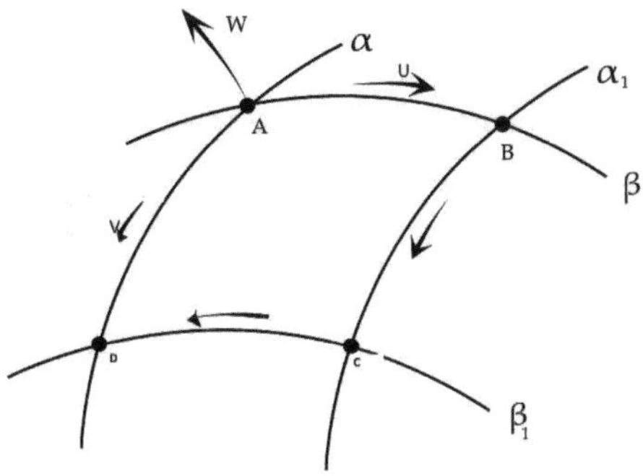

Fig. 7.4 $\mathscr{C}(\mathbf{u}, \mathbf{v})\mathbf{w}$ defines the curvature of the manifold at A, with respect to the affine connection Γ_μ

Such a topological definition of curvature is more general than the Riemann definition of curvature which is specific for a local metric geometry.

In particular for a generic reference frame $\{\mathbf{e}_\mu\}$ defined in the quantum space-time, the components of the affine connection defined above are

$$\Gamma_\rho \overset{def}{=} g_{\mu\sigma}\Gamma^\sigma_{\mu\rho} = (log\sqrt{-g})_{,\rho}, \tag{7.3}$$

and the components of the curvature of the quantum space-time are

$$\mathscr{C}(\mathbf{e}_\mu, \mathbf{e}_\nu)\mathbf{e}_\rho = [D_\mu, D_\nu]\mathbf{e}_\rho = \left(\partial_\mu \Gamma_\nu - \partial_\nu \Gamma_\mu + [\Gamma_\mu, \Gamma_\nu]\right)\mathbf{e}_\rho. \tag{7.4}$$

Note that with the condition $D_\rho g_{\mu\nu} = 0$, we obtain the same components of the Riemann tensor (2.7) as described in Chap. 2, now written for the metric $g_{\mu\nu} = g(\mathbf{e}_\mu, \mathbf{e}_\nu) =< \mathbf{e}_\mu, \mathbf{e}_\nu >$.

Since in the quantum space-time there are no observers, the metric does not work as an observational instrument. In this case, we use the diffeomorphism covariant derivative (7.1), whose components are defined by an *adjoint representation of the group of diffeomorphisms of the space-time*. As we may readily see, the end result is that the above expression of curvature coincides with the Riemann definition of curvature.

$$\mathscr{C}(\mathbf{e}_\mu, \mathbf{e}_\nu)\mathbf{e}_\rho = R_{\mu\nu\rho\sigma}\mathbf{e}^\sigma = \left((\Gamma_{\nu\rho\mu,\sigma} - \Gamma_{\nu\sigma\mu,\rho}) + (\Gamma^\tau_{\nu\rho}\Gamma_{\mu\sigma\tau} - \Gamma^\tau_{\mu\rho}\Gamma_{\nu\sigma\tau})\right)\mathbf{e}^\sigma. \tag{7.5}$$

The specification of a diffeomorphism covariant affine connection for Γ_ρ is relevant to keep the compatibility with Einstein's principle of general covariance.

Therefore, and similarly to what happens with gauge theory, there is a Lie algebra associated with the quantum gravitational field. This is the Lie algebra of the quantum group of diffeomorphisms, acting on the quantum space-time. Consequently the principle of general covariance is also implicit in the quantum version of Einstein's gravitational field. In particular, we have seen that the determinant of the Jacobian matrix of that Lie algebra has been identified with the term $\sqrt{-g}$ in the quantum version of the Einstein-Hilbert action.

Chapter 8
The Geometry of the Standard Model

Abstract Starting from the hypothesis that the Standard Model is complete, we may obtain a set of equations capable of describing everything: Einstein's quantum gravity, plus the three Yang Mills fields, plus a set of geometrical conditions describing something new: A relativistic version of the Poincaré conjecture within the quantum domain.

The past conjectured existence of an extra-dimensional compact spaces in classical higher dimensional gravitation of the non-Abelian Kaluza-Klein theory has posed a real problem for physics, because these compact spaces are inconsistent with the observed mirror (or chiral) symmetry at the electroweak scale of unification. Another related problem is that the higher dimensional gravitational prevents eventually should be quantized by some yet undefined process. Finally, the presence of **G** assumed in higher dimensional gravitational theories the renormalization of the gravitational field. We have seen that these problems in a higher dimensional unification of the fundamental interactions may be solved by use of the renormalizable perturbations of the Minkowski metric by the massless spin-2 quantum field, obtaining the same gravitation in the sense of Einstein, except by the use of the Newtonian paradigm.

The first and the most well known among the higher dimensional theories is the five-dimensional Kaluza-Klein theory originally proposed by Theodor Kaluza in 1918, describing unification of the then-known fundamental interactions (actually only gravitation and electromagnetism) in a five-dimensional space-time [142, 143]. In 1926 Felix Klein suggested that the Kaluza theory would make sense, provided it was defined at the quantum scale [144]. The mathematical definition of the Kaluza-Klein theory was criticized by Einstein and Bergman in 1938 [133, 145]. However, in 1963 the Kaluza-Klein theory was extend to include all three gauge fields in a classical higher dimensional space-time. Although it was mathematically correct, the non-Abelian Kaluza-Klein theory as it became known was eventually dismissed in 1984 it was found that the theory was incompatible with the chiral symmetries of the gauge fields which were defined in the quantum domain. We shall return to this subject in Chap. 10.

M. D. Maia, E. M. Monte, *Geometry at the Quantum Scale*,
https://doi.org/10.1007/978-3-031-61197-1_8

In the present chapter we will pursue a similar unification program, with the difference that Einstein's gravitational field is already defined at the quantum scale, as a consequence of the renormalizable perturbations of the Minkowski metric by the spin-2 massless quantum field, leading to the same Einstein-Hilbert action, now defined at the quantum scale. As a consequence we will be in conditions to reexamine the non-Abelian Kaluza-Klein program, as defined in the 11-dimensional quantum space, with a geometry defined by the four submanifolds associated with the four fundamental interactions defined at the quantum scale. The Standard Model of Unification will provide the experimental basis for the unification program, to be called *The Geometry of the Standard Model.*

8.1 The Differential Equations of the Standard Model

In the following we will apply the Nash-Greene theorems that properly handle the local inclusion of differentiable Riemannian manifolds with positive or negative metric signatures within a higher dimensional space-time [147, 148], applied to the 11-dimensional quantum space geometry. With such theorems we also eliminate the limitations and impediments of the previously known Janet-Cartan theorems. Therefore, we are able to define geometry of the Standard Model of Interactions as applied to the 11-dimensional quantum space. In doing so, we consider two separate situations.

(a) Initially we will consider that the the Standard Model in its present form as being complete, in the sense that there are only four fundamental interactions, the three gauge forces and the gravitational force, all represented by an 11-dimensional at Riemannian manifold, with a vanishing Riemannian curvature $R_{ABCD} = 0$
(b) Secondly, if a new fundamental force in nature is eventually found and experimentally confirmed, then such new force will be represented by a new and non-vanishing curvature component of the 11-dimensional space, or equivalently, it can be represented by a sub-manifold embedded in a higher-dimensional (beyond 11-dimensions) at space[1]. In any case, we will start with the hypothesis (a).

Denote the components of the Riemann tensor of the 11-dimensional quantum space (actually it is an 11-dimensional quantum space-time) by ${}^{11}R_{ABCD}$. In particular, for practical purposes we may define a basis of the 11-dimensional quantum space, as separated into two parts: Four of these vectors denoted by $\mathbf{e}_\mu$ are tangent to the four-dimensional quantum space-time, and the remaining seven

[1] The 11-dimensional space At the time of this writing (June 2025) the Fermilab announced the conclusion of the g-2 experiment [149]. Therefore, we assume that the 11-dimensional space is flat.

components denoted by η^a are orthogonal to the four-dimensional quantum space-time:

$$\begin{cases} \text{space-time tangent vectors } \mathbf{e}_\mu, \quad \text{with } \mu,\ \nu\ldots = 1,\ 2,\ 3,\ 4\ \ g_{\mu\nu} =< \mathbf{e}_\mu, \mathbf{e}_\nu > \\ \text{space-time normal vectors } \boldsymbol{\eta}_a, \quad \text{with } a\, b\ \ldots = 5,\ 6\ \ldots,\ 11,\ \ g_{ab} =< \boldsymbol{\eta}_a, \boldsymbol{\eta}_b > \\ \qquad \text{since they are orthogonal vectors, then } g_{\mu a} =< \mathbf{e}_\mu, \boldsymbol{\eta}_a >= 0. \end{cases}$$

Using such basis the equations ${}^{11}R_{ABCD} = 0$ may be decomposed in three sets of equations that are similar to the equations in classical differential geometry of submanifolds, known as the Gauss-Codazzi-Ricci equations, here applied to quantum geometry. The derivation of these equations can be found in a very detailed exposition, although using different and a somewhat difficult notation for physical interpretation in [24]. An adaptation of these equations for a higher dimensional classical Riemannian geometry, in the context of classical cosmology, can also be found in [150, 151, 163]

$$ {}^{11}R_{ABCD} = \begin{cases} {}^4R_{\mu\nu\rho\sigma} - 2g^{ab}k_{\alpha[\gamma a}k_{\delta]\beta b} = 0 \text{ (Gauss)} \\ k_{\alpha[\gamma b;\delta]} - g^{mn}A_{[\gamma mb}k_{\alpha\delta]n} = 0 \text{ (Codazzi)} \\ -2g^{mn}A_{[\gamma ma}A_{\delta]nb} - 2A_{[\gamma ab;\delta]} - g^{mn}k_{[\gamma ma}k_{\delta]nb} = 0 \text{ (Ricci)} \end{cases} \tag{8.1}$$

where we have denoted

$$\begin{aligned} &g_{\mu\nu} =< \mathbf{e}_\mu, \mathbf{e}_\nu > && \text{The first fundamental form of the quantum space-time,} \\ &k_{\mu\nu a} =< \mathbf{e}_\mu, \eta_{a,\nu} > && \text{The second fundamental form of the quantum space-time,} \\ &A_{\mu ab} =< \eta_{a,\mu}, \eta_{b,\mu} > && \text{The third Fundamental form of the quantum space-time.} \end{aligned} \tag{8.2}$$

In differential geometry these equations are well known to define the conditions for existence of submanifolds of a differentiable higher dimensional manifold (here in particular for a flat manifold). The difference is in their physical interpretations: The metric components $g_{\mu\nu}$ are called *the first fundamental form of geometry,* while in high energy physics they are the components of the gravitational potential, the components of the *third fundamental form of geometry* $A_{\mu ab}$ in high energy physics are the components of the gauge potentials, and finally, *the second fundamental form, also known as the extrinsic curvature*, does not correspond to any previously known potential of a force field, although it has some relevant physical implications described below. Interestingly the second fundamental form has been applied to a five-dimensional space-time extension of Einstein's classical gravitational theory, after postulating an expression for the second fundamental form [152, 153].

8.2 The Relativistic Poincaré Conjecture

The earliest account for the study of submanifolds of a flat manifold dates back to the nineteenth century [154], followed by Ricci and Levi-Civita [105]. Perhaps the most detailed derivation of the abovementioned equations, including some applications, although using a different notation can be found in [24].

The resulting equations are known in differential geometry as the Gauss-Codazzi-Ricci equations given by (8.1), describing the relationship between the three fundamental forms of geometry. Here we have changed Eisenhart's notation to make it easier to identify the corresponding potential functions in physics. Note that the Gauss-Codazzi-Ricci equations do not replace the dynamical equations. Rather, they just give a geometric relationship between physically significant quantities. While in classical space-times this identification has been considered in most studies of classical gravitation to be irrelevant [138], in the quantum geometry without hierarchies between the interactions, these equations become an essential complement to the dynamical equations. This is so because they provide the necessary condition for the compatibility between these physical quantities defined in the same space. We may say in other words that these equations provide the compatibility of the interaction between gravitation and the gauge fields.

One interesting fact is that while the first and third fundamental forms of geometry are easily identified with the components of gravitational field $g_{\mu\nu}$ and of the gauge potentials $A^a_{\mu b}$, respectively, the second fundamental form $\kappa_{\mu\nu a}$ apparently does not have any known physical function. In the following we explore the analogy between the second fundamental form and a relativistic generalization of the Ricci flow, as a possible short-range massive spin-2 field representing dark matter at the quantum scale.

In order to understand the abovementioned analogy, let us recall a conjecture made by H. Poincaré in 1904, stating that a *closed, smooth, simply connected 3-manifold is diffeomorphic to a sphere* S^3 [155, 156]. Poincaré never proved his conjecture. However, in 1982 Richard Hamilton proposed a geometrical equation that he called the Ricci flow, starting from the expression for the Ricci tensor

$$R_{\mu\nu} = (log\sqrt{g})_{,\mu\nu} - \Gamma^{\rho}_{\mu\nu;\rho} + \Gamma^{\sigma}_{\mu\rho}\Gamma^{\rho}_{\nu\sigma} - \Gamma^{\rho}_{\mu\nu}(log\sqrt{g})_{,\rho}.$$

Using geodesic coordinates, the terms involving the Christoffel symbols in the above expression disappear so that the expression of the corresponding Ricci scalar simplifies to

$$R = g^{\mu\nu}R_{\mu\nu} = \nabla^2(log\sqrt{g}).$$

Denoting $log\sqrt{g} = u$ *and assuming that* u *satisfy Fourier's heat equation*, he obtained $\nabla^2 u = \frac{\partial u}{\partial t}$. After canceling the above trace and replacing the time variable t by a coordinate y orthogonal to a two-dimensional surface embedded in a two-dimensional space, Hamilton obtained what he called the *Ricci Flow* [157]:

$$R_{\mu\nu} = \frac{1}{2}\frac{\partial g_{\mu\nu}}{\partial y}. \tag{8.3}$$

As we see, the Ricci tensor is described by the variation of the metric of the 2-surface, along a geodesic orthogonal to the surface of the manifold, hence the suggestive name "Ricci Flow."

In 2002, G. Perelman used the above Ricci flow expression to show that a two-dimensional smooth closed surface in a three-dimensional Riemannian manifold may be smoothly deformed into a two-dimensional sphere, just by following the heat flow lines [158, 159]. A detailed proof of the Poincaré theorem can be found in [156]. Some interesting notes on this subject can also be found in [160].

The Poincaré conjecture is actually a description of a physical phenomenon that is part of our everyday experience in classical physics, describing a deformation of a surface into another, while keeping its basic topological characteristics. Such deformation can be explained by the application of a local external force on the orthogonal lines of a plastic (or deformable) closed surface to obtain a spherical surface, something like we do with a play dough.

The question that poses now is: Is the Ricci flow something fundamental? To answer, consider our everyday observation of the deformation of the surface of the Earth, resulting from the Newtonian gravitational pull of the Moon. Another common example is the Faraday electric induction that produces a current in two nearby electric conductors, inducing a local deformation of the electromagnetic flux.

The Poincaré conjecture can also be applied at the astrophysical scale of events to explain the observed nearly flat velocity rotation curves of stars in spiral galaxies. There must exist something like a non-visible matter (or dark matter), such that it corrects the centripetal force concentrated around the galaxy nucleus. Likewise we may use the Poincaré conjecture to explain the formation of stars out of the cosmic dust of the nebula, but this is more elaborate.

Since the Einstein gravitational field of General Relativity is a spin-2 field characterized by the space-time curvature, then, unlike in the Newtonian gravitation, if we place two massive particles out there, say in a Lagrange point between the Earth and the Sun, nothing should happen simply because two particles do not generate a relativistic force (contrarily to what happens in the Newtonian gravitation). On the other hand if we place three particles near the same point, then we may observe some gravitational force, but not gravitational waves. Actually, in order to obtain a genuine relativistic gravitational force, we will need a minimum of four particles, capable of generating a relativistic "center of mass", thus forming another center of mass. As we see, after a (long) while, we should expect to observe the formation of gravitating object, like the formation of a star or even black hole in the space-time. However the Ricci flow is not consistent with Einstein's equations. To see this, write Einstein's equations as

$$R_{\mu\nu} = 8\pi G\left(T_{\mu\nu} - \frac{1}{2}T g_{\mu\nu}\right).$$

Then for an arbitrary source $T_{\mu\nu}$, it follows that the Ricci flow (8.3) condition implies that Einstein's equations become a linear equation

$$\frac{\partial g_{\mu\nu}}{\partial y} = -16\pi G\left(T_{\mu\nu} - \frac{1}{2}T g_{\mu\nu}\right),$$

which means that the Ricci flow is not compatible with Einstein's Relativity. However, it suggests a possible existence of a similar relativistic version of the Ricci flow, consistent with Einstein's equations.

8.3 The Nash Flow

In the derivation of the quantum gravitational field from a massless spin-2 field, the hierarchy of the fundamental interactions was also eliminated because the Newtonian constant was not included, mainly because it is incompatible with the renormalization of Einstein's gravitational field. Therefore, assuming that the four known fundamental forces in Nature (i.e. assuming that there is no additional fundamental force.) in the Standard Model, then the resulting set of equations summarized in ${}^{11}R_{ABCD} = 0$ can be seen as a geometrical representation of the Standard Model of Interactions. In fact, the resulting Gauss-Codazzi-Ricci equations of differential geometry equations summarized by (8.1) represent a relationship between the quantum gravitational force represented by gravitational potential (the first fundamental form) $g_{\mu\nu}$ and the quantum gauge forces represented by the gauge potentials $A^a_{\mu b}$ (the third fundamental form) of the 11-dimensional quantum space-time geometry.

In order to understand the physical meaning of the second fundamental form, we need to understand the concept of *Lie derivative*. In Chap. 3 we have seen that by dragging a vector field $\mathbf{w}$ along a closed geodesic loop on a manifold, starting and ending at the same point p, we obtain the curvature of the manifold at p expressed by the exponential of the loop integral (3.3). Now, suppose that after traveling the *infinitesimal* length of the loop $d\xi$, the vector $\mathbf{w}$ gives up the trip and decides to go back to the same initial point *using the same original path* in the reverse direction, just to find that the transported vector $\mathbf{w}$ now called $\mathbf{w}'$ *does not necessarily coincide with the original one*, except if the manifold is flat. The difference $\mathbf{w}' - \mathbf{w}$ divided by the variation of the parameter interval between the beginning and end of the trip, calculated at the initial point, is called the *Lie derivative* of $\mathbf{w}$ along the direction $d\xi$, defined as

$$\pounds_\xi \mathbf{w} = -\left(\frac{\mathbf{w}' - \mathbf{w}}{\Delta t}\right)\rfloor_{\Delta t \to 0}. \tag{8.4}$$

The minus sign in the expression (8.4) specifies that a reverse direction was taken with respect to some previously chosen path. *The vector* $\mathbf{w}'$ *is often called the Lie*

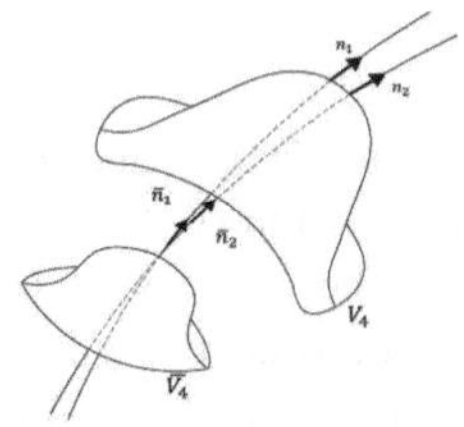

Fig. 8.1 A local Lie deformation of the original manifold

transport of **w**. Mathematically, the Lie derivative can be made in a more formal way (e.g., as in [161]). An intuitive physical interpretation can be found in [162].

We may now apply the Lie transport to the tangent vectors of a reference frame $\{\bar{\mathbf{e}}_\mu\}$, with respect to normal directions $\bar{\eta}_\mathbf{a}$, and we obtain a *Lie transported reference frame*

$$\{\mathbf{e}_\mu, \boldsymbol{\eta}_a\} \to \{\bar{\mathbf{e}}_\mu + \pounds_{\bar{\eta}}\bar{\mathbf{e}}_\mu, \bar{\boldsymbol{\eta}}_a\} \tag{8.5}$$

defining *local Lie deformation* of a neighborhood in the four-dimensional quantum space-time, relative to orthogonal directions of some gauge field, as tentatively drawn in the Fig. 8.1.

Applying such deformation to all components of the three fundamental forms of an embedded geometry in the 11-dimensional quantum space, we obtain the fundamental forms of the locally deformed geometry

$$\text{The fundamental forms of the deformed geometry} \left\{ \begin{array}{l} g_{\mu\nu}(x,y) = \bar{g}_{\mu\nu} - 2y^a\bar{k}_{\mu\nu a} + y^a y^b[\bar{g}^{\alpha\beta}\bar{k}_{\mu\alpha a}\bar{k}_{\nu\beta b} + g^{cd}\bar{A}_{\mu ca}\bar{A}_{\nu db}] \\ g_{\mu a}(x,y) = y^b\bar{A}_{\mu ab} \\ g_{ab}(x,y) = \bar{g}_{ab} \\ k_{\mu\nu a}(x,y) = \bar{k}_{\mu\nu a} - [\bar{g}^{\alpha\beta}\bar{k}_{\mu\alpha a}\bar{k}_{\nu\beta b} - g^{cd}y^b\bar{A}_{\mu ca}\bar{A}_{\nu db}], \\ A_{\mu ab}(x,y) = \bar{A}_{\mu ab}(x) \end{array} \right. \tag{8.6}$$

where $\bar{g}_{\mu\nu}$, $\bar{k}_{\mu\nu a}$, and $\bar{A}_{\mu ab}$ denote the fundamental forms of the original geometry.

One interesting relation is obtained by calculating the derivative of the first fundamental for (the metric) $g_{\mu\nu}$ with respect to y^a, and comparing the result with the expression of the second fundamental form $k_{\mu\nu a}$, we obtain

$$k_{\mu\nu a} = -\frac{1}{2}\frac{\partial g_{\mu\nu}}{\partial y^a}. \tag{8.7}$$

As it is transparent, the right-hand side of the above expression is the same expression as the Ricci flow, with the exception that instead of the derivative with respect to the time variable, we have the derivatives with respect to the coordinates in the orthogonal direction η_a of the gauge fields in the quantum space-time. It is interesting to the extent that in the left-hand side, we have the expression of the extrinsic curvature of the space-time relative to each orthogonal direction η_a. Therefore, the above expression describes a generalized foliation of the space-time, whose leaves are four-dimensional deformed space-times, defined for each extra dimension of the quantum space-time.

The expression (8.7) is suggestive: It tells that the extrinsic curvatures of the space-time describe the local deformations of the original space-time along the orthogonal directions. This is a generalization of the Ricci flow to the quantum domain. In particular it tells that it is possible to form a spherical singularity induced by the presence of gauge fields, eventually forming a black hole. The following picture is not very different from the classical Poincaré conjecture. It gives an intuitive idea of what we call a Nash flow, showing how a closed space-time region may be continuously deformed into a sphere, perhaps to form a star or a black hole (Fig. 8.2).

Note that contrarily to the Ricci flow, the Nash flow is not limited to a single and always increasing time direction. As we shall see, it is possible to reverse the flow so that the black sphere may also deform into a new space-time shape, suggesting a possible geometric explanation to the black hole entropy.

In an attempt to give an example of the physical implications of the Nash flow deformation of the space-time at the quantum scale, consider that we have only the quantum electromagnetic field. Supposing that the four-dimensional gravitation and the electromagnetic field are complete, and in a state of equilibrium (no additional force is assumed), then the equations for such a simple model is given by the vanishing of the five-dimensional curvature tensor

$$^5R_{ABCD} = 0. \tag{8.8}$$

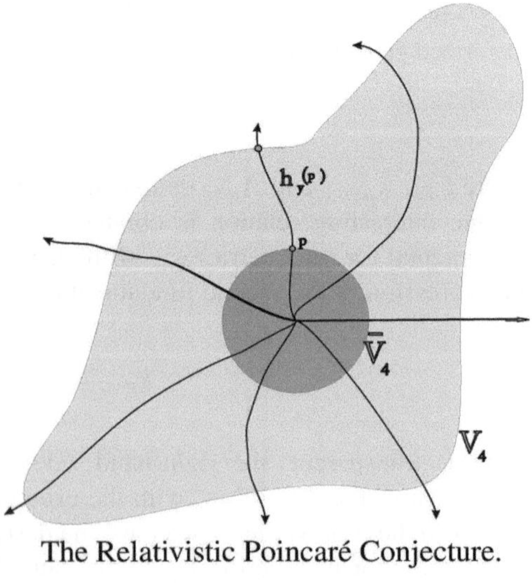

Fig. 8.2 The Relativistic Poincaré Conjecture

In this example the deformation of the four-dimensional space-time occurs only along a single fifth dimension, given by the Lie transport of the tangent reference frame along the single quantum extra dimension η_5, leading to the equations

$$^5R_{\mu\nu} = R_{\mu\nu} + \frac{\partial k_{\mu\nu}}{\partial y} - 2k_{\mu\rho}k^{\rho}_{\nu} + hh_{\mu\nu} = 0, \tag{8.9}$$

where in this example $k_{\mu\nu 5} = k_{\mu\nu}$, $R_{\mu\nu}$ is the Ricci tensor of the 4-dimensional quantum space-time, R denotes the Ricci scalar and where $H = g^{\mu\nu}k_{\mu\nu}$ is the mean curvature of the space-time, leading to the equations

$$R_{\mu\nu} - \frac{1}{2}Rg_{\mu\nu} - Q_{\mu\nu} = 0 \tag{8.10}$$

$$k_{\mu;\rho}^{\ \rho} - H_{,\mu} = 0 \tag{8.11}$$

where we have denoted $k_{\mu\nu 5} = k_{\mu\nu}$, $H = g^{\mu\nu}k_{\mu\nu}$ is the mean curvature of the space-time with respect to the 5th dimension and

$$Q_{\mu\nu} = g^{\rho\sigma}k_{\mu\rho}k_{\nu\sigma} - k_{\mu\nu}h - \frac{1}{2}\left(K^2 - H^2\right)g_{\mu\nu}, \tag{8.12}$$

where $K^2 = k^{\mu\nu}k_{\mu\nu}$. It follows that $Q_{\mu\nu}$ is a conserved quantity with respect to the 4-dimensional covariant derivative

$$Q^{\mu\nu}{}_{;\nu} = 0. \tag{8.13}$$

stating that the tensor $Q_{\mu\nu}$ is an observable seen by the classical space-time observer. A suggestive name for $Q_{\mu\nu}$ would be the *quantum space-time deformation tensor induced by the electromagnetic field.* We may consider a continuous sequence of such deformations along a single orthogonal direction, as defining a *one-parameter group of diffeomorphism*, along the curve tangent to the unit normal vector η.

8.4 Dark Matter Flows

The example of the previous section suggests that the Nash flow can be seen as a local deformation of the space-time, which is given by the above conserved tensor $Q_{\mu\nu}$ in the quantum space-time. A four-dimensional observer may interpret such an effect as the result of some additional force or energy that has a *local* effect on the space-time geometry. However it is not described by the space-time metric, which is a long-range (nonlocal) field. Then $Q_{\mu\nu}$ must be a local short-range gravitational effect that is not contemplated by Einstein's gravitation theory except perhaps in its Newtonian limit. Since here we are talking about quantum gravity, then such deformation must be related to quantum field theory or more specifically as a spin-2 quantum field. Indeed, since the massless spin-2 field was already identified as the generator of a renormalizable perturbation of the Minkowski metric to produce Einstein's massless gravitational field, then the only remaining explanation for the above local deformation of the space-time is that it describes a short-range spin-2 field.

As we have seen the existence of a massive spin-2 field has been considered before, and it is at the root of a somewhat sensitive subject, the hypothesis that the massless Einstein gravitation (General Relativity) can in principle be derived from the zero mass limit of massive gravitation. The problem with such interpretation follows from David Boulware's conclusion that the presence of the mentioned such zero mass limit implies in the emergence of ghost particles in the (quantum) space-time [88]. We have already discussed this subject in Chaps. 4 and 5, concluding that there is no evidence for the existence for the zero mass limit in nature.

As we have explained, starting hypothesis in the construction of the Fierz-Pauli Lagrangian was that the condition $h = 0$ was assumed so that the Fierz-Pauli field has the correct number of degrees of freedom, so that the minimal energy condition always applies. However the massless case is primary because the Fierz-Pauli field is already in its ground state without any further condition, and it is precisely that condition that warrants the Einstein gravitational field to be a long range field. The presence of a nonzero mass makes the massive gravitational field a short-range field. The problem lies precisely in the supposition that the long-range field comes from the zero mass limit of the short-range one. As we have already discussed at length in Chap. 5, such a limit does not exist in nature: Massive particles remain massive, and massless particles remain massless.

Massive gravity as a spin-2 short-range field can be identified geometrically by the above Nash deformation tensor of the space-time geometry resulting from the presence of gauge fields, provided it is independent of the massless gravitational field. This is so because there is no such thing as a zero mass limit within the spectrum of the Casimir mass operator. However, dark matter may be explained as the local deformation of the quantum geometry of the space-time, related by the presence of the gauge fields. Thus, let us look again at the Fierz-Pauli Lagrangian with a nonzero mass term.

$$\mathscr{L}_{Fierz-Pauli} = \frac{1}{4}[h_{,\mu}h^{,\mu} - h_{\nu\rho,\mu}h^{\nu\rho,\mu} - 2h_{\mu\nu}{}^{,\mu}h^{,\nu} + 2h_{\nu\rho,\mu}h^{\mu\rho,\nu} - U]. \qquad (8.14)$$

Since we still have the traceless condition $h = 0$, and we still have a spin-2 field, then it follows that $h_{\mu\nu}$ has five nonzero solutions in its state of minimal potential energy, defined by the minimal energy condition which we may denote as $h_{(a)\mu\nu}$, for $a = 1, ..., 5$

$$\frac{\partial U}{\partial h_{(a)\mu\nu}} = \mu^2(h_a^{\mu\nu} - h_a\eta^{\mu\nu}) = 0, \quad \text{for} \quad \mu \neq 0,$$

satisfying the field Eqs. (8.14)

$$(\square^2 - {\mu_a}^2)h_{(a)\mu\nu} = 0, \tag{8.15}$$

which obviously describe a set of short-range massive wave equations, with decaying amplitude, orthogonal to the space-time, and with a different mass m_a. Consequently, the above massive Fierz-Pauli equations describe a periodic deformation of the space-time geometry with decaying amplitudes, much in the sense of the Poincaré conjecture, adapted to the Nash smoothing condition. In other words, the classic Nash smoothing condition is the same as we call the extrinsic curvature for the quantum space-time.

$$k_{\mu\nu a} = h_{(a)\mu\nu} = -\frac{1}{2}\frac{\partial g_{\mu\nu}}{\partial y^a}, \tag{8.16}$$

describing *a decaying quantum fluctuations* of the space-time. This is somewhat similar to the quantum fluctuations associated with loop quantum gravity. Indeed these fluctuations are periodic fluctuations produced by a gauge field. The difference with loop quantum gravity is that here quantum gravity has been already defined by the long-range renormalizable perturbations of the massless spin-2 Fierz-Pauli field, while the deformation obeys a massive, and hence local (decaying) wave equation. Note also that the extrinsic curvature is a fundamental concept in geometry necessary to solve the Eqs. (8.1) as a standard differential equations, dispensing with the rather demanding analytic condition required by the Janet-Cartan embedding theorem. Finally it should also be noted that instead of the thermodynamical argument as used by the Ricci flow, here we have a local deformation of the space-time geometry imposed by the presence of gauge forces at the quantum domain.

From the mathematical point of view, the second fundamental form $k_{\mu\nu a}$ is independent of the metric (the first fundamental form). The presence of the second fundamental form is part of the system of equations that define a hypersurface, and it is an essential condition for the existence and uniqueness of solution of the Euler-Lagrange equations. This is so because mass is a truly discrete set of numbers, which implies that for each value of the coordinate y^a we have an independent component $k_{\mu\nu a}$ producing a different solution of the Eqs. (8.15). This strongly suggests that massless and massive gravitation are truly distinct and independent dynamical fields at the quantum scale, in the sense that one cannot be taken as being the zero mass limit of the other.

Chapter 9
The Higgs Roulette

Abstract The completion of the equations of the Standard Model of Interactions at the quantum scale still depends on the solution of one specific problem: How the Higgs assigns the correct mass to the particles of the standard Model at the same time that it agrees with the O'Raifeartaigh's theorem? Here we show how such a fundamental problem can be solved with the aid of the rotating Higgs mechanism, a device that tells how the masses of each particle of the Standard Model are properly assigned.

If Einstein's gravitation is quantized, then at the quantum scale there must exist a four-dimensional quantum version of the space-time with the same local tangent Minkowski space-time, whose metric is invariant under the Poincaré symmetry. As it happens, the masses of the quantum particles of the Standard Model are also determined by the mass operator of the Poincaré group. Consequently, the symmetry mixing problem becomes more acute at the quantum scale, and it must be somehow solved, before anything else. As we have detailed in Chap. 6, this is a fundamental problem of high energy physics that is specific for the Poincaré group. The symmetry mixing problem emerged in the 1960s from the O'Raifeartaigh's theorem showing that it is not possible to combine the Poincaré group of symmetries with the symmetry groups of gauge fields in any arbitrary way, under the penalty that all particles with the same internal spin must also have the same (squared) masses. This is not true as it can be readily seen on the by Standard Model (Fig. 1.3).

Note however that the symmetry mixing problem is specific for the Poincaré group, simply because the mass spectrum of fundamental particles is defined by that group. Therefore it applies to any other unification model that uses the Poincaré symmetry as it is. As we have seen in Chap. 1, the Euler-Lagrange equations of the electromagnetic, weak, and strong nuclear forces rest at the very basis of the Standard Model. If we change the fundamental system of physics, like, for example, in string theory, then the Poincaré symmetry may also be changed, although at the end, say in a lower energy limit, the Poincaré symmetry must be recovered. However, as it has been proven in the replacement of the Poincaré group is not a trivial task.

M. D. Maia, E. M. Monte, *Geometry at the Quantum Scale*,
https://doi.org/10.1007/978-3-031-61197-1_9

On the other hand, the Standard Model lacks a formal definition or something that could be called a Theory of the Standard Model, perhaps something that may start with the definition of a Lagrangian and its symmetries. The first step on that direction was the introduction of a mechanism capable to assign to each particle of the Standard Model its Poincaré-defined mass, presented by Peter Higgs and others back in 1964 [134, 164, 165]. Such a mechanism proposed the existence of a particle (the Higgs particle), such that it interacts with all the other particles of the Standard Model, in such a way that when the Higgs particle reaches its state of minimum energy (its vacuum state), it assigns the correct mass of the other particle. The Higgs particle was experimentally found in 2012 at the LHC. However, the Higgs mechanism is not very explicit on how it works. Indeed, since these masses must come from the mass spectrum of the Casimir operator of the Poincaré group, then it must also contain at least implicitly a solution of the symmetry mixing problem. Here we show that this is true: The existence of the Higgs Particle solves the symmetry mixing problem.

Since the Standard Model is invariant under the Poincaré group, all its particle masses must come from the spectrum of eigenvalues of the Casimir mass operator (similarly, the orbital spins must come from the spectrum of the Casimir spin operator of the Poincaré group) [116, 117, 167]. It is here that lies the problem: For any arbitrary combination of the Poincaré group with the gauge groups of symmetries, the predicted masses do not agree with observed ones. More specifically, the symmetry mixing problem shows that if these particles have the same internal spin, then they should also necessarily have the same masses in accordance with (6.5).

The problem deserves full attention because we cannot get anywhere without an explanation on how to differentiate the particle masses. The problem received a great deal of attention with several proposals as, for example, in [110, 118–122], among many others.

The Coleman-Mandula theorem of 1967 asked something like: What would be the most general symmetry compatible with the observed mass differences? The answer came in the form of radical changes including changes in the Lie algebra structure together with the concept of elementary particles. String theory is one possible example of such a formal structure where instead of point particles the fundamental structure would be composed of one-dimensional strings, which could be quantized in a Poincaré invariant space-time with 26 dimensions or in a ten-dimensional space-time with the use of the supersymmetric paradigm was previously tried with supergravity, without success. Half century later, we still do not have a definitive solution of the symmetry mixing problem based on the Poincaré symmetry.

In order to understand how the Higgs mechanism works, the starting point is Wigner's theorem on the classification of the unitary and irreducible representations of the Poincaré group. That theorem says that the masses of the particles of the Standard Model belong to a very special set of positive numbers which populate the set of eigenvalues of the mass Casimir operator defined by $C^2 = P_\mu P^\mu$ where

P_μ denotes the Poincaré translation operator described in Chaps. 1 and 2. As it happens, the spectrum of eigenvalues of that operator is a special set composed of *discrete and unevenly spaced natural numbers*. Unlike the set of integer or half integer numbers set of spins of the particles, the mass spectrum is very discrete with points separated by unpredicted distances. These numbers are determined only by high energy accelerator experiments. We simply do not know how such a set appears or are generated. Perhaps it represents a genuine gift of nature.

9.1 The God Particle

The mechanism proposed by Higgs required the existence of a new fundamental particle (the Higgs particle) associated with a new symmetry group, such that when that new symmetry ceases to exist (or to use the field theoretic terminology "when that symmetry is spontaneously broken"), the mechanism "assigns" a mass to a particle of the Standard Model, in the spectrum of the Casimir mass operator [166].

In order to detail the Higgs mechanism we need to write its Lagrangian $\mathscr{L}_{Higgs}$, containing a potential energy and its own group of symmetry, called the Higgs symmetry. The Higgs proposition was that the Higgs particle, in its vacuum state (the state of minimum energy), interacts with any other particle of the Standard Model, in such a way that no energy is transferred from the Higgs to the other particle. However, when the Higgs symmetry ceases to exist, without any additional intervention of the observer (*a spontaneously broken symmetry*), then the Higgs particle points to the spectrum of the mass Casimir operator $C^2 = P^\mu P_\mu$, indicating the mass of the said particle. In short, in its vacuum state, the Higgs particle acts as a pointer indicating the mass of the particles of the standard model. People tend to ask where these masses came from? They came from something that nature gave us, or to use Lederman's own words, God gave to us in the form of the eigenvalues of the mass Casimir operator.

In order to understand the Higgs mechanism, consider a generalization of the Klein-Gordon equation given by

$$\Box^2\phi + \sum_{n=0}^{\infty} \Lambda_n \phi^n = 0. \tag{9.1}$$

From this equation we may obtain many interesting equations like, for example, the sine-Gordon equations.

In particular, choosing $\Lambda_0 = 0,\ \Lambda_1 = \mu^2 < 0,\ \Lambda_2 = 0,\ \Lambda_3 = \lambda/3! > 0$ and assuming that all other coefficients are zero, we obtain the equation

$$(\Box^2 + \mu^2)\phi + \frac{\lambda}{3!}\phi^3 = 0,$$

which is the Euler-Lagrange equation resulting from the Lagrangian

$$\mathscr{L}_{Higgs} = \frac{1}{2}\eta^{\mu\nu}\phi_{,\mu}\phi_{,\nu} - U(\phi),$$

where

$$U(\phi) = \frac{1}{2}\mu^2\phi^2 + \frac{\lambda}{4!}\phi^4$$

plays the role of the potential energy for ϕ.

As always the measurement of ϕ is made in the state of minimum potential energy, characterized by

$$\frac{\partial U}{\partial \phi} = \mu^2\phi + \frac{\lambda}{3!}\phi^3 = 0$$

whose possible solutions are given by

$$\left(\mu^2 + \frac{\lambda}{3!}\phi^2\right)\phi = 0, \quad \Rightarrow \quad \begin{cases} \phi = 0 \text{ or} \\ \phi = \pm\sqrt{\frac{-6\mu^2}{\lambda}} = \pm a. \end{cases} \tag{9.2}$$

Since the Higgs is a quantum particle, then the potential energy

$$U(\phi) = U_0 = \text{constant}$$

defines a quartic two-dimensional surface in the space in the quantum domain. However the above simple two-dimensional model does not show all features of the Higgs field which is in fact a two-dimensional manifold defined in the complete 11-dimensional quantum space of the Standard Model. For example, the solution $\phi = 0$ is usually discarded under the allegation that such a solution does not represent a stable vacuum: In other words, given any small perturbation, say during a measurement, the vacuum solution $\phi = 0$ will not remain in that position, and consequently that particular vacuum solution is not considered a true vacuum.

On the other hand, the other two solutions of (9.1) lead to the generalized Klein-Gordon equation

$$(\Box^2 - \mu^2)\phi = \frac{-\lambda}{3!}\phi^3$$

Showing that the Higgs field acquire the same squared mass $m^2 = -\mu^2$ for any two nontrivial vacuum positions $a = \pm\sqrt{\frac{6m^2}{\lambda}}$.

Actually there is a vacuum symmetry, which in this example is given by the discrete group of translation from $\phi = a$ to $\phi = -a$. Thus the spontaneous symmetry braking means that the choice of $\phi = a$ or of $\phi = -a$ is irrelevant, because the resulting mass-squared value $m^2 = -\mu^2$ is given by nature, independently of the observer. Hence, the expression that the symmetry of the Higgs vacuum $a \leftrightarrow -a$ is *spontaneously broken*. The spontaneity is a reference to the fact that it happens without any intervention of the observer. However, the above model is too simple to show the full picture of the whole set of symmetries that is involved in the Higgs mechanism.

9.2 The Rotating Higgs

In order to understand the idea of the *spontaneous symmetry breaking*, we need to work with the more realistic case of the electroweak field interacting with the Higgs field, described by the Higgs Lagrangian, written in the quantum space. As we have seen the electroweak is described by the symmetry product $U(1) \times SU(2)$ whose respective internal geometries are the one-dimensional circle in which the symmetry corresponds to a rotation of a circle whose points are mapped by the complex set, and $SU(2)$ is isomorphic to $SO(3)$ which corresponds to the rotations on a 2-dimensional spherical surface whose points are mapped by quaternions. Therefore, we end up with complex quaternions that are represented by the complex 2×2 Pauli matrices σ_μ. Thus the electroweak field may be represented by 2×2 complex matrices, and the Higgs field is represented by a complex column, sometimes referred to as the "Higgs doublet"

$$\phi = \begin{pmatrix} \phi_1 \\ \phi_2 \end{pmatrix}. \quad \phi^\dagger = (\phi_1^*, \phi_2^*)\,. \tag{9.3}$$

The interacting Electroweak-Higgs field is represented by Lagrangian

$$\mathscr{L}_{EW-Higgs} = \left[\underbrace{-\frac{1}{4} tr\, F^{\mu\nu} F_{\mu\nu}}_{Electroweak} + \underbrace{(D^\mu \phi)^\dagger (D_\mu \phi)}_{Mixing\ term} - \underbrace{\mu^2(\phi^\dagger \phi) + \lambda(\phi^\dagger \phi)^2}_{Higgs} \right], \quad \lambda > 0, \tag{9.4}$$

where the first term represents the Lagrangian of the gauge quantum field A_μ and whose field strength (or curvature tensor) is $F_{\mu\nu} = [D_\mu, D_\nu]$, where $D_\mu = \partial_\mu + A_\mu$ is the gauge covariant derivative associated with the gauge potential A_μ. The second term represents the interaction between the gauge field and the Higgs field. Finally, the last term represents the Higgs Lagrangian.

Denoting the Higgs potential by $U(\phi_1, \phi_2)$, the minimum potential energy condition is now given by

$$\frac{\partial U}{\partial \phi} = -\mu^2 \phi^\dagger + 2\lambda(\phi^\dagger \phi)\phi^\dagger = 0. \tag{9.5}$$

The above Lagrangian (9.4) is invariant under the action of the Poincaré group acting on the internal quantum space-time, together with the electroweak symmetry $U(1) \times SU(2)$, acting on a four-dimensional internal space. The Higgs potential has the form usually described as the "Mexican hat," which is a fourth order surface [1] drawn in the hypersurface drawn in the 11-dimensional quantum space.

Similarly to what happened in the two-dimensional example, the solution $\phi = 0$ of (9.5) is traditionally excluded, under the allegations that the vacuum at the top is *not stable*. For this reason, the $\phi = 0$ vacuum is sometimes referred to a *false vacuum*. The truly stable vacuum states are the points located at the bottom circumference shown the Fig. 9.1 of the next page. Actually, it is quite intuitive to see in the above figure that any small perturbation of a particle located at the point $\phi = 0$ will make it drop down to the true vacuum states located at the bottom circumference.

More recently, in the case of the electroweak symmetry $U(1) \times SU(2)$, it has been conjectured that these instabilities may be produced by the "quantum fluctuations of the Higgs potential" [187]. However such interpretation may jeopardize the whole purpose of the Higgs mechanism, which is to assign the masses of the other particles. Indeed, these quantum fluctuations would also apply to the point $\phi = 0$, thus dismantling the whole purpose of the Higgs mechanism.

Actually, the elimination of the vacuum solution at $\phi = 0$ is not a problem of the instability of the Higgs potential. Rather, in order to fulfill its purpose as a fundamental particle of the Standard Model, indicating the masses of the other particles of that model, the Higgs must also have a *global symmetry* $SO(2)_{\text{Global}}$ in addition to its own local $SO(2)_{Local}$ symmetry. The $SO(2)_{Global}$ symmetry characterizes the Higgs potential as a surface of revolution around the axis $\phi = 0$ in the Fig. 9.1 below. Since the position $\phi = 0$ coincides with the point where the axis of revolution would cross that surface, then that point cannot belong to the surface, simply because the tangent vectors to the Higgs surface are not uniquely defined. Said in other words, at that point the Higgs potential would be a nonregular surface, and consequently it must be removed.[2]

We must keep in mind that we are talking about a surface defined at the quantum domain. Therefore the definition of how small would be the diameter of the hole through the revolution axis is really not precise. As in every measure in quantum theory, it will depend on Heisenberg's uncertainty relations, so that the mentioned quantum fluctuations of the Higgs vacuum are admissible within the limits imposed by the Heisenberg relations. Beyond those limits there are no instabilities to consider near the point $\phi = 0$, simply because that point does not exist in the surface.

---

[1] This means a power 4 term in the equation. It can be easily checked by just drawing a straight line crossing the hat. The maximum number of points crossing the line tells the order of the surface.

[2] In a surface of revolution the axis cannot have a common point with the axis of revolution. A regular surface means that the tangent vectors in the neighborhood of a point can be well defined in all directions.

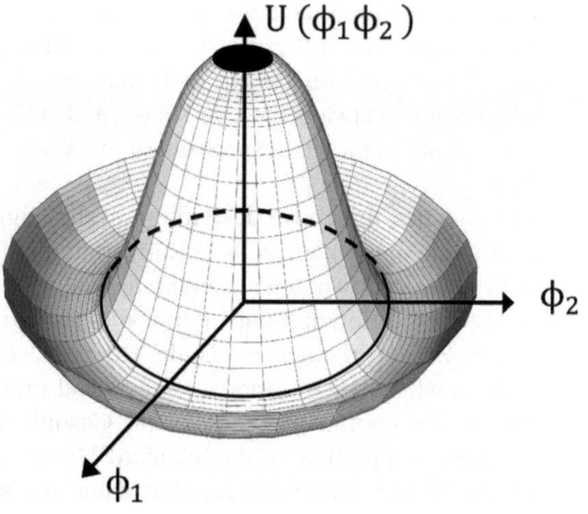

Fig. 9.1 The Higgs potential in the electroweak field

Consequently, the only true and stable vacuum states of the Higgs are the points located at the bottom circumference of the rotating Higgs potential.

The intuitiveness of the Higgs potential as a surface of revolution is also supported by the thought experiment in which the Higgs particle falls into a rotating potential surface, as pictured in Fig. 9.1, where the dark circle at the top represents the hole of the Higgs potential, whose diameter is as small as permitted by Heisenberg's relations. Then in the same thought experiment the reader may imagine that the falling Higgs gets in contact with the rotating surface, remaining there until the centrifugal force provided by the rotation of the surface is overpowered by the weight of the Higgs particle.[3] Then, only after the global rotation of the surface ceases completely (or to use the physics language, "when the global rotation symmetry $SO(2)$ is broken"), the Higgs particle drops to the bottom circumference where it remains at one fixed rest position, in its state of minimal energy. At that rest point the observers may read the position at the bottom circumference, say, the point a, where the squared Higgs mass m_a can be measured. Only after the rotation stops, the position (or slots) where the Higgs particle rests, it points to a mass of a particle of the Standard Model, provided we draw a 1:1 correspondence between the mass spectrum of the Casimir mass operator (normally are plotted as points in a straight line), as points into a fixed circumference close to the Higgs vacuum circumference (as shown in Fig. 9.1), so that the position where the Higgs particle stops also acts as a pointer to the mass spectrum of the Casimir mass operator. Note that this happens only after the Higgs global symmetry breaks down, so that its position also points to the mass of a particle of the Standard Model.

[3] Remembering that in such an experiment we are assuming the existence of a quantum gravitational field.

As the reader may have already noted, the rotating Higgs potential resembles Casino roulette. The difference is that the playing ball is the Higgs particle, and instead of money, nature plays with masses: At the end of each run, the Higgs particle maintains always the same mass, and at the same time it points to a specific mass belonging to the mass spectrum of the Mass Casimir Operator as shown in the Fig. 9.2:

Note also that the above description of the Higgs potential as a rotation surface is in complete agreement with O'Raifeartaigh's theorem: When the global Higgs symmetry breaks down, the squared mass value of the Higgs is always the same: If a and b are two different points of the bottom circumference where the Higgs stops, then $m_a^2 = m_b^2$, or in other words, the Higgs mass is always the same. However, at the position where it stops, there is also a local symmetry breaking, where the Higgs vacuum position points to a mass of the Casimir mass operator, thus assigning the correct mass to a particle of the Standard Model *within the context of the Poincaré group*. The rotating Higgs potential explains not only the Higgs mass but also how the Higgs determines the masses of particles in the Standard Model.

As indicated in the Fig. 9.2, the experimentally determined mass spectrum represented by the horizontal line is 1:1 mapped on color coded slots in a fixed position around the rim of the vacuum Higgs circumference. These slots are also marked (all gray) in the rotating roulette, without any numbers. When the roulette stops (meaning that the $SO(2)$ global symmetry is broken), the position of the slot where Higgs particle rested points to a mass of a particle of the Standard Model.

Finally it should be noted also that there are in fact two symmetry breakings, one after the other: Only after the *global* $SO(2)$, symmetry is broken (when the Higgs particle stops at the gray slot in the figure), and the *local* $SO(2)$ symmetry is also broken (when the roulette finally stops), so that the Higgs particle points to the winning mass.

As a final remark we note that the Higgs Roulette is defined at the quantum domain, including Einstein's quantum gravitational field without hierarchies (there-

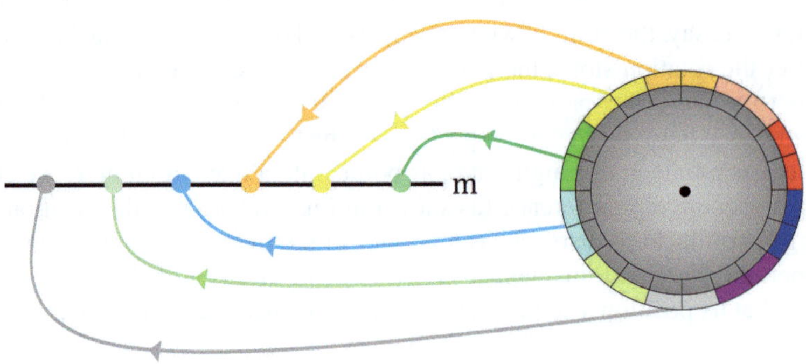

Fig. 9.2 In the Higgs Roulette, the mass spectrum is one-to-one mapped into the outer rim of the rotating surface so that when the global rotation symmetry stops, the position of the Higgs points to a natural particle mass on the table so that nature always wins

fore without **G**). Assuming that the Standard Model is complete, then we have an 11-dimensional geometry with zero curvature, in which all known interactions have comparable strengths. This led us to a new quantum cosmology to study the evolution of the entire universe, with the participation of the whole set of known fundamental interactions and fundamental particles.

Chapter 10
Quantum Black Holes

Abstract Hawking's concept of mini black holes could in principle tell something about the quantum of gravitation. However, such understanding is hampered by the conjectured information loss problem. Here, in the context of the renormalizable Einstein's Gravitation, we show that the information loss problem in fact does not exist: Particles, virtual or otherwise, never fall inside a singularity, simply because a singularity is a forbidden region of the quantum space-time. Hence any physical information always remains within the space-time, although near a black hole the information cannot be read, due to the presence of the strong gravitational field.

10.1 Singularities and Their Horizons

The first formal solution of Einstein's equations showing the existence of a singular boundary was given by Karl Schwarzschild in 1916, just after Einstein presented his theory. The Schwarzschild's solution represents spherically symmetric static vacuum solution of Einstein's equations in General Relativity, specified by Schwarzschild's line element written in spherical coordinates (r, θ, ϕ, t) shown in Chap. 2) as

$$ds^2 = -\left(1 - \frac{2\mathbf{G}M}{r}\right) dt^2 + \left(1 - \frac{2\mathbf{G}M}{r}\right)^{-1} dr^2 + r^2(d\theta^2 + sin^2\theta d\phi^2), \tag{10.1}$$

where M denotes an integration constant, which is usually interpreted as the Newtonian mass, resulting from taking the Newtonian limit of that solution. Such a boundary condition is correct only in the theory of General Relativity which is a theory derived from Einstein's generalization of Newtonian gravitation. The

M. D. Maia, E. M. Monte, *Geometry at the Quantum Scale*,
https://doi.org/10.1007/978-3-031-61197-1_10

spherically symmetric shell with radius $r = r_{Schw} = 2\mathbf{G}M$ is called the horizon of Schwarzschild's singularity.[1]

However, the presence of **G** in (10.1) means that Schwarzschild's singularity contains a mass, in the Newtonian sense. That is, that mass is composed of mass in the Newtonian sense, composed by atoms and their aggregates. This is so because **G** is determined by torsion balances using lumps of atoms (often referred to as ordinary matter). As such, the presence of **G** in (10.1) excludes the possible action of any other internal force of quantum nature that may exist within the atomic nuclei which are not detected by a torsion balance. Such exclusion suggests that General Relativity is essentially a classical theory of gravitation, in the sense that it is not affected or that it does not affect quantum forces. Such conclusion appears to be in full agreement with the 't Hooft-Veltman proof that General Relativity is a non-renormalizable theory of gravitation, essentially as a consequence of the presence of **G**.

The first definition of a space-time singularity as seen from a distant observer is due to J.R. Oppenheimer: *A singularity is a region where the space-time may cease to exist* [168]. On the other hand, Finkelstein in 1958 was more specific when he stated that *the Schwarzschild singularity has a boundary called the horizon, which is a perfect unidirectional membrane, in which the causality principle applies only in one direction. Accordingly, a particle falling all the way from a point at infinity to inside the horizon disappears forever, never to be seen again* [169].

The first physical evidence of a singularity in space-time occurred in 1964, during the astronomical observation of a source of X-rays in our galaxy: It was noted that those rays were coming from a star denominated Cygnus X-1, which was orbiting an *unknown dark object*, near the center of our galaxy. Today, many astronomical observations of such dark objects have been reported, justifying their designation as *black holes in (classical) space-time*.

Allegedly the name black hole was firstly used by John Wheeler, suggesting that a singularity is a hole in space-time. From such designation came the idea that objects in space-time may in principle fall inside that hole by gravitational attraction. However, no explanations were given concerning the conservation laws and how exactly the trajectories of the falling particles should be treated around that singularity as usually we do when solving similar boundary value problems in classical mechanics. Note also that Einstein's assumption and later verification that light is also attracted by the gravitational field tell that a gravitational-electromagnetic interaction exists even though light has no mass in the Newtonian sense.

In 1960 M. Kruskal found that Schwarzschild's singularity horizon could be eliminated simply by using different coordinates [170] (shown in Fig. 10.1). The Kruskal solution is topologically distinct from Schwarzschild's solution in the sense that the former solution is geodesically complete, while the later solution is not.

---

[1] Most people working in General Relativity do not explicitly write the Newtonian Gravitational constant, by assuming the Planck convention telling us that $c = \hbar = \mathbf{G} = 1$ hold universally.

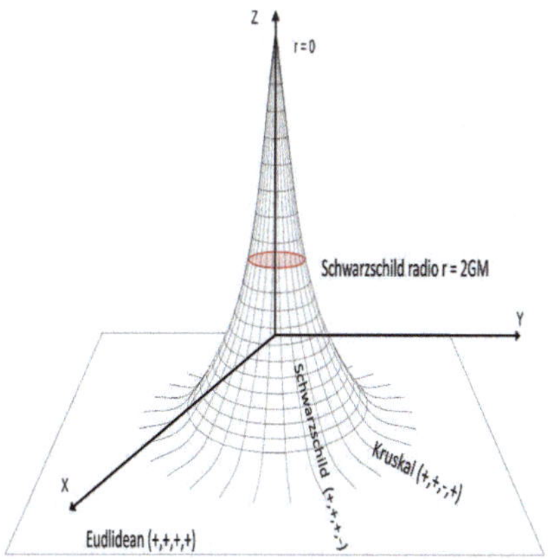

Fig. 10.1 Superposing the Schwarzschild, Kruskal, and Euclidean Space-times

In other words, the concept of singularity (black holes) is space-time has some relevant topological implications to the structure of the physical space-time, which in principle should be taken into considerations. In fact, since the principle of general covariance allows all possible coordinates, then the Kruskal solution result could in principle be obtained as maximal extension of Schwarzschild's solution. While in the Schwarzschild case the geodesics of a radially falling particle ends at the black hole horizon, the Kruskal solution shows that the particle falls all the way from infinity to the origin, with an intriguing detail: The definition of time changes when crossing the horizon.

10.2 Mini Black Holes

In the 1970s Stephen Hawking suggested that *at the quantum scale there should exist some high-density quantum packets of energy that he described as mini black holes* [171, 172]. These mini black holes would be similar to the classical black holes, in the sense that they are also singularities of the quantum space-time. However, as a quantum energy-packet, those mini black holes can be associated with a massless spin-2 particle defined within the context of quantum gravity. In the later development of his theory, Hawking assumed that these mini black holes could have a submanifold embedded within the singularity, described as a "memory cache" in which the supposed lost information was kept and could be recovered after the "singularity evaporates."

In accordance with Hawking's description, the mini black holes would have originated in the universe just after the big bang and that they could still exist and be observed today. Hawking suggested that similarly to what happens in high energy quantum physics, the strong gravitational field of those mini black holes would also create pairs of virtual particles and antiparticles. While the virtual particles are attracted by and falls inside the mini black hole horizon, the virtual antiparticles flee away from the horizon. These virtual antiparticles eventually collide with other nearby particles, thus producing an endless sequence of creation and annihilation of particle pairs. Assuming that such a process is renormalizable, in principle a classical observer will perceive some kind of short-lived radiation similar to the Hawking-Bekenstein black hole entropy defined in General Relativity by

$$S_{BH} = \frac{c^3 A_{BH}}{4\mathbf{G}\hbar}, \tag{10.2}$$

where A_{BH} denotes the area of the black hole limited by its horizon [173–175]. As we can see that the entropy increases proportionally to the increase of its area. Since in nature the entropy is always increasing, then its area is also always increasing, preventing the observation of the quantum status of the particle that felt inside the black hole horizon. Consequently, Hawking's mini black hole proposition implies in a conceptual problem that was never solved: *If a quantum particle falls inside the black hole horizon, then the observers will not see any information about its quantum status*, preventing the verification of the conservation of the probability, a stronghold of quantum field theory.

As we have seen in Chap. 5, the Einstein-Hilbert action can be obtained from the renormalizable perturbations of the Minkowski metric by a series of powers of the massless spin-2 field. The resulting Einstein-Hilbert action is identical to Einstein's gravitation without any reference to the Newtonian gravitational constant, which is replaced by an arbitrary coupling constant κ. However, the metric $g_{\mu\nu}$ is no longer a measuring instrument compatible with quantum theory. This is so because we the observers remain classical. For a quantum space-time the instrument of observation must be replaced by a diffeomorphism covariant affine connection defined by the derivative of the determinant of the Jacobian matrix of the group of diffeomorphisms of the classical space-time. Consequently, the solutions of Einstein's equations in quantum gravity are the same as in classical gravitation, consistently associated with the principle of general covariance of classical relativistic gravitation, with the only notable difference: The Newtonian coupling constant has to be changed to a coupling constant κ that is compatible with the renormalization of Einstein's gravitation defined by the Einstein-Hilbert principle of General Covariance.

With such providences, we may modify the Hawking-Bekenstein expression for a black hole entropy at the quantum scale, by just replacing the Newtonian constant **G** by the coupling constant κ, obtaining what we may call the *renormalizable black hole entropy*.

$$S_{BH} = \frac{c^3 A_{BH}}{4\kappa\hbar} \tag{10.3}$$

in which Hawking's mini black hole radiation can in principle be detected at the laboratory, perhaps even at the presently available TEV scale energy. Some estimates tell that such radiation may be copiously produced in the laboratory and that they evaporate very rapidly [176, 177].

The original version of the mini black hole hypothesis used the Wheeler-de Witt model for canonical quantum gravity. That model is similar to the classic Hamilton-Jacobi theory, in the sense that time needs to be recovered in its classical limit. However, since the evaporation of the mini black hole produces an infinite sequence of creations and annihilations of virtual particle pairs, the comparison of those processes would in principle depend on how time is recovered. Perhaps to avoid further discussions related to the definition of time between the in-falling state and the time in which the information would be recovered, in 2004 Hawking changed the model of quantum gravity to the *Euclidean quantum gravity*, in which the Minkowski space-time geometry is transformed into a Euclidean geometry using a "Wick rotation" (essentially replacing the real time coordinate by an imaginary time coordinate at the quantum scale, so that the space-time metric ds^2 looks like the Euclidean metric) [178, 179]. With such a modification, Hawking suggested that the supposed lost information remained inside a three-dimensional submanifold, like some kind of memory cache embedded inside the black hole horizon, so that the information can be recovered after the black hole evaporates.

A less known advantage of the Euclidean quantum gravity over the Wheeler-de Witt theory resides in the fact that it makes it possible to consider the classic Janet-Cartan theorem on the existence of submanifolds embedded in an Euclidean space [180, 181], thus guaranteeing the existence of the three-dimensional memory cache manifold as properly embedded in the Euclidean space-time.[2] Of course, the Euclidean model still depends on the proof of existence and stability required by the memory cache inside the black hole singularity. Only with such stability, the supposed lost information could be in principle recovered after the evaporation of the black hole [182].

Note that Fig. 10.1 shows that Kruskal space-time is geodesically complete, while the Schwarzschild space-time is not. Therefore the difference between these solution are physically significant. In the Kruskal space-time, time and space are interchanged across the horizon.[3] The Nash theorem is a relevant generalization of the Janet-Cartan theorem, allowing the embedding of differentiable manifolds [147], while the Greene theorem extends Nash's theorem to space-times with nonpositive metrics [148].

On the other hand, for a quantum space-time the observational conditions are different from the classical space-time. For example, Hawking's mini black holes cannot be characterized as a classical singularity. This is so because the observers cannot access the quantum domain, much less carying a metric wand to the quantum domain. For example, the classical Kerr black hole solution of Einstein's equations of General Relativity is characterized by its mass, charge, and angular momentum.

---

[2] In fact the Janet Cartan theorem also requires analytic manifolds, perhaps an additional requirement for Hawking's memory cache.

[3] The existence of a submanifold of a space-time is not trivial. Back in 1975 we have suggested the application of the Nash's theorem to Hawking's hypothesis on the existence of a memory cash as a submanifold of a mini black hole without appealing to the Euclidean quantum gravity.

However at the quantum scale, while the same singularity must be characterized by its spin and its energy, for example, as compared with the Planck energy, or perhaps as compared with an effective quantum scale [183, 190].

10.3 Quantum Black Holes

The presently existing plethora of models for quantum gravity is a symptom that there is something wrong with Einstein's General Relativity. In the previous chapter we have identified two relevant causes for such a problem: the presence of the Newtonian gravitational constant in Einstein's equations and the neglect of the hyperbolic characteristics of Einstein's equations. With the solution of these two problems, we have seen that the Einstein-Hilbert principle and Einstein's equations can be derived directly from a massless spin-2 quantum field, without any reference to the Newtonian theory of gravitation, making it possible for any classical observer to observe and measure Einstein's gravitational field, with a single difference: the replacement of **G** by a coupling constant compatible with the definition of a gravitational field that is compatible with quantum field theory. For those who are still concerned with a canonical formulation of quantum gravity, we have pointed out in in Chap. 3 that Dirac's formulation of quantum gravity is the only option consistent with the hyperbolic characteristic of Einstein's equations. It implies that there exists a four-dimensional quantum space-time, defined within the quantum domain, whose metric remains as the gravitational potential, just like in the classical space-time solution of the same Einstein equations. The difference is that the observers are and remain classical, so that a different method of observation for the quantum space-time is required. Here we kept the same metric geometry as in the classical General Relativity, representing the potential if the gravitational field. However, the observational method is different, using the diffeomorphism invariant affine connection Γ_ρ. Using such the affine connection in the quantum space-time, we have described a general covariant, hierarchically free, and renormalizable quantum gravity characterized by the same Einstein-Hilbert action, with the same metric as representing the gravitational potential, whose curvature tensor components $R_{\mu\nu\rho\sigma}$ are the same as the curvature defined by the metric Riemann tensor, producing the same Riemannian curvature.

With the above summary and precautions we may now return to Hawking's mini black hole problem, now defined in the four-dimensional quantum space-time defined as a massless spin-2 field solution of Einstein's equations, using the general covariant affine connection defined by the renormalizable perturbations of the Minkowski metric by the massless spin-2 quantum field, obtaining essentially the same Einstein equations, with the exception that instead of the Newtonian gravitational constant which is proper to the Newtonian classic and classical gravitation, we used a coupling constant κ to be consistent with sources of quantum gravity, using the general covariant affine connection Γ_ρ.

Returning to Hawking's description, of the mini black holes defined at the quantum scale, the quantum virtual antiparticles flee away from the mini black hole horizon, thus interacting with other nearby particle and radiation, producing what he called the evaporation of the mini black hole. However the fate of the horizon of the singularity is not too clear. Would it also disappear with the evaporation, thus forming what is called a naked singularity, something like black hole without a horizon?

From the point of view of geometry, the horizon of a classical black hole can be characterized as the last three-dimensional surface that may be identified by a distant observer *using photons*. In the case of quantum gravity it is not clear how to define a naked singularity because, without hierarchies, all types of high energy probes are available, so that we really do not know about naked singularities at the quantum scale.

However, within quantum geometry, at our disposal, we may talk about the horizon of a mini Schwarzschild black hole, with an observational problem: A particle coming from a large distance (depending of course on what is meant by large distances at the quantum scale. By default, we choose it to be something larger than the Planck scale of lengths) may have an infinite self-energy, and this is not what is observed by the classical observers. The concept of Schwarzschild black hole may not exist at the quantum scale, due to the lack of hierarchy between all fundamental interactions, so that more realistically, black holes may be endowed with a rotation, like a mini Kerr black hole. It is always useful to remember that at the quantum scale we may not appeal to the Newtonian limit to determine the integration constants for solutions of Einstein's equations, as we normally do in classical gravitation. Consequently, a particle falling into a mini black hole will be necessarily falling at an angle, and consequently it tends to follow an orbit around the mini black hole horizon.

More realistically, we are talking about a quantum Kerr black hole, endowed with an angular momentum, so that the particles falling on it become trapped in a closed orbit by the strong gravitational field. In this case, any information that was supposed to be lost inside the black hole horizon will become visible to the distant observer after the evaporation of the quantum black hole.

The Fig. 10.2 shows an orbit that is so close to the horizon that the classical observer may think that the quantum information was lost. In fact it was never lost, but it remains there hidden by the strong gravitational field of the mini black hole. However if the black hole evaporates, then the supposed lost information may be recovered.

Reciprocally the evaporation of the mini black hole may be understood as a consequence of the increase of the area of the horizon by its overpopulation by trapped orbits, thus producing an increased entropy. In the simple case of Schwarzschild's black hole, the entropy is given by

$$E_{Schw} = \frac{\mathscr{A}}{4k}, \quad (\text{with } c = \hbar = 1), \tag{10.4}$$

Fig. 10.2 Trapped orbits around a quantum mini black hole

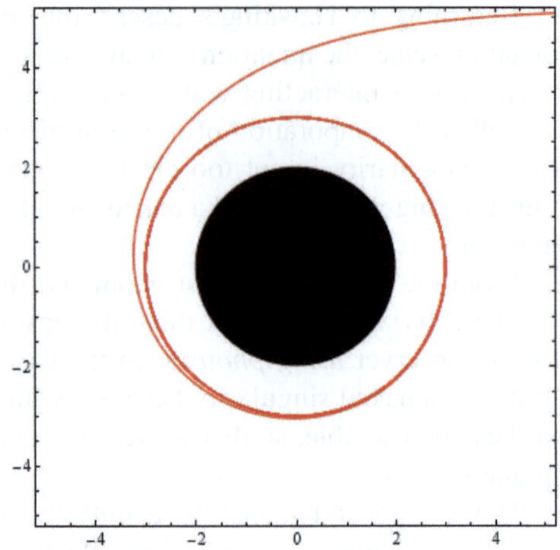

where $\mathscr{A}$ denotes the area of the mini black hole. As we see, the analogy with thermodynamics is almost perfect in the sense that evaporation of the quantum black hole occurs at expense of the expansion of the area of the horizon. We say almost perfect because the expansion of the horizon may also prevent the observer to see the horizon after the evaporation.

10.4 The Geometry of the Standard Model

The detailed structure of the Lagrangian of the Standard Model was explicitly calculated by M. Veltman, displaying a formidable mathematical structure based on the observational contents of that model [184, 185]. Here we will describe the Einstein-Hilbert Lagrangian of the Standard Model, including only the four fundamental forces and their interactions.

Using the recent result of the g-2 experiment at the Fermilab, the Standard Model can be seen as the simplest combination of the known four forces of nature in complete harmony, represented by a flat 11-dimensional topological manifold, expressed by the vanishing of the curvature tensor $R_{ABCD} = 0$. This leads to a system of equations involving the three fundamental forms of differential geometry: The first differential form of geometry (or the gravitational potential with components $g_{\mu\nu}$), seen as a solution of the renormalizable Einstein equations defined at the quantum scale, the second fundamental forms of geometry, seen as the gauge potential with components A_{μ}, defined as solutions of the Yang-Mills equations, also defined at the quantum scale, and finally, the third fundamental form with components $k_{\mu\nu a}$, which, differently from the two previously mentioned forms, do not represent a solution of a known dynamical equation. Rather, the second fundamental form represents the Nash Flow (8.7), which was seen to be a relativistic generalization of the Ricci flow, describing deformations of the space-time geometry along the extra dimensions defined by the gauge potentials. The second fundamental form satisfies a condition that was proposed by Nash, regarded here as a generalization of the Ricci flow, where instead of a time variation of the space-time metric, it is defined by the variation of the space-time metric, with respect to the seven extra dimensions associated with the gauge forces defined at the quantum scale. Such interpretation is interesting because it provides a physical meaning to the second fundamental form, similar to the geometrical interpretation of the Poincaré conjecture.

The resulting set of Eqs. (8.1) was originally derived and understood within the geometry of submanifolds, and here they are interpreted as a set of equations involving the four known fundamental interactions. Even more interesting is the fact that these equations were also shown to be derivable from the Lagrangian functional in a higher dimensional classical space-time. This conclusion was reached in 1963 in an exercise proposed in a course of lectures to generalize the old five-dimensional Kaluza-Klein theory (described in [142–144]), now defined by a higher dimensional Einstein-Hilbert variational principle [146].

The assumption that the abovementioned Kaluza-Klein theory was of classical (non-quantum) nature had the consequence that the theory used the same classical space-time of the observers. Consequently, as it was found in 1984 the theory was dismissed because it was inconsistent with the observed chiral symmetry in the four-dimensional space-time. In detail, that inconsistency was essentially due to the fact that the higher dimensional geometry was regarded as a classical geometry applied to a classical space-time. To compensate the difference with quantum geometry, the extra dimensions were required to be very small and compact, which in turn was later found to imply in the existence of large masses at the electroweak range of applications, which is not compatible with the observations. In addition, the Kaluza-Klein theory based on the same Einstein-Hilbert principle would still need to be quantized.

As we have emphasized along several instances of the present notes, it is possible to derive a renormalizable perturbation of the Minkowski metric by the massless spin-2 quantum field $h_{\mu\nu}$, leading to a renormalizable quantum gravitational field equation in the same sense of the Einstein-Hilbert variational principle. The only remark to be added is that the observers remain classical, as defined in their classical Riemann space-time. This meant that the space-time metric cannot be used as a measuring instrument to be applicable to the quantum scale. Instead of the metric the classical observers can use a new covariant vector affine connection Γ_ρ which is defined by the group of diffeomorphisms of the classical space-time. The result is a completely consistent higher dimensional Einstein-Hilbert Lagrangian defined in the 11-dimensional manifold characterized within the quantum scale of events, excluding the Newtonian gravitational constant.

With the above precautions and in analogy with the classic non-Abelian Kaluza-Klein Lagrangian described in Chap. 7, we were able to construct the *Quantum Kaluza-Klein Theory* involving the four fundamental interactions of the Standard Model, defined in the 11-dimensional quantum manifold, with the Einstein-Hilbert

Lagrangian functional defined in a $(4 + N)$-dimensional space.[4] to write the more general Lagrangian (The upper parenthesis on the left denotes the dimension of the quantum space.)

$$\mathscr{L}_{\text{The Standard Model}} = {}^{(4+7)}R\sqrt{G} = {}^{4}R\sqrt{-g} + \frac{1}{4} tr\, F_{\mu\nu}F^{\mu\nu}\sqrt{-g}, \tag{10.5}$$

where we have used the fact that in general a Lagrangian functional is not defined in space-times, but rather in the *configuration space* of the physical system, which must also include not only the main field components but also their momenta as independent variables. In principle, we also need to add boundary conditions, required to guarantee the existence and uniqueness of the solutions of the resulting Euler-Lagrange equations.

Note that the configuration space must have a sufficient number of dimensions to accommodate all gauge fields plus gravitation. That is, we need to directly sum the three spaces associated with the respective dimensions 1, 2, and 4 for the gauge fields, together with a four-dimensional space-time with a tangent Minkowski space-time. So that the configuration space has a minimum of 11 dimensions. Such simplest space can be obtained by the local direct sum of the gauge field space, plus the space-time defined at the quantum domain. It is interesting to notice that this is the same number of dimensions for M-theory.

Since the gauge fields are defined locally at each point of the space-time, the construction of a fiber bundle within the quantum domain is similar to the construction of fiber bundles on classical geometry, with the relevant difference that the base space of the fiber bundle is the quantum space-time and where each fiber is the direct sum of the tangent Minkowski space-time, plus the 7-dimensional internal projective space $\mathscr{B}_N$ where the gauge groups act, similar to the construction described, for example, in [186]. The base space is the four-dimensional quantum space-time, not to be confused with the classical space-time. Thus, for example, we may define the fiber bundle structure of the above 11-dimensional Standard Model as

$$(V_4, \pi, V_4 \times \mathscr{B}_N), \tag{10.6}$$

where V_4 is the quantum space-time, π denotes the projection map, and $\mathscr{B}_N$ denotes the total space, represented in the Fig. 10.3 below.

This is the simplest construction obtained by what is called a trivial fiber bundle, in which the total space (the configuration space in mechanics) looks like a box locally described by the Cartesian product of the quantum four-dimensional space-time base space, times a copy of the seven-dimensional fibers, which are

[4] Here, at the risk of excessive generality, we are extending the $4 + 7$ dimensions to $4 + N$, so as to include the possibility of new results in experimental high energy physics, including new forces or new "things" (as, e.g., the Nash flow), which may require additional dimensions. For the present Standard Model, just take $N = 7$.

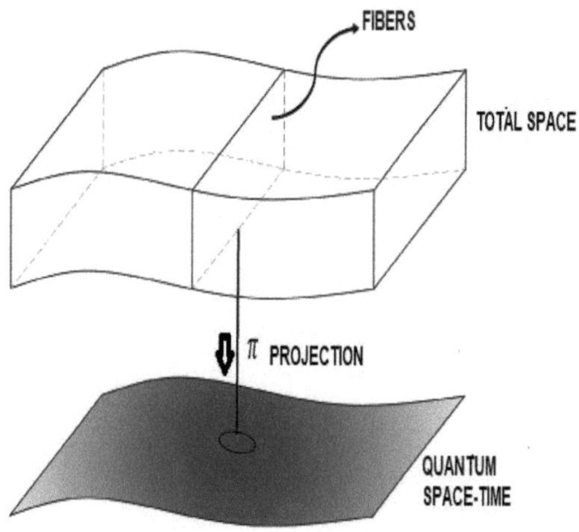

Fig. 10.3 The bundle of the standard model

all isomorphic to each other. These fibers are here defined by the direct sum of the quantum spaces where the respective Lie algebras of the groups $U(1)$, $SU(2)$, and $SU(3)$ act. Thus, each fiber is composed of the direct sum of the respective gauge spaces, a circle S_1, a sphere S_2, and the four-dimensional space where the exceptional group associated with the $SU(3)$ acts, leading to the total fiber space denoted by $\mathscr{B}_7$ given by (10.6).

The Lagrangian describing the standard model is a map from that 11-dimensional space into the real line. Following the results obtained for the non-Abelian Kaluza-Klein theory proposed in [146], the action principle defined in the 11-dimensional space decomposes into the four-dimensional Lagrangian for the gravitational field in the four-dimensional space-time, plus the four-dimensional Yang-Mills Lagrangian for the Standard Model:

$$\int {}^{(4+7)}R\sqrt{-{}^{(4+7)}g}dv^{(4+7)} = \int R\sqrt{-g}dv + \int \frac{1}{4}tr\,F_{\mu\nu}F^{\mu\nu}\sqrt{-g}dv, \qquad (10.7)$$

where all integrals in the right-hand side are defined over four-dimensional volumes.

The expression (10.7) is intriguing: It suggests a generalization of the 5-manifold theory in the sense of Donaldson's theorem [21, 22]. Accordingly, it tells that the reduction to four dimensions is not an exclusive property of the Yang-Mills theory. With the inclusion of the four-dimensional quantum space-time, the Donaldson theorem applies to all known fundamental interactions.

Concluding Remarks

The recent measurements in astronomy, ranging from the infrared to microwave frequencies, suggest that cosmology is facing a conceptual crisis. Sadly, we realize that we know only about 5% of our universe. On the other hand, at the quantum scale of high energy physics, we have fared better, with the establishment of the Standard Model of Fundamental Interactions, whose strength lies in the fact that it is mostly based on the experimental evidences. This book is about the establishment of a geometrical model for the Standard Model.

We start with the assumption that the currently observed universe cannot be described by a classical solution of Einstein's equations, involving only ordinary atomic matter. Rather, the current observations tell that ordinary classical matter contribute to only to less than 5% of the observed energy contente of the universe. On the other hand we find that the Standard Model of Interactions needs a geometrical language so as to include gravitation as a fundamental force of nature. The result may be called Quantum Cosmology, a discipline involving all known fundamental interactions hopefully explaining the remaining 95% of the universe.

Why a geometrical language? The geometrical language is natural because it has proven to be the necessary language to the description of classical physics, although it has not been frequently used in quantum physics. If we want to understand the universe in terms of quantum physics, then we need also to understand the role of geometry at the quantum scale. More to the point, we need to understand the geometry that is behind the Standard Model of Interactions. It took several years of hard work to combine the essential mathematical and physical development, to the present development of differential topology in mathematics, to conclude that a geometry of the Standard Model is given by an 11-dimensional Einstein-Hilbert action. The hard part was to convince ourselves that some definitions postulated by 1915 need to be updated in face of the present experimental high energy physics, astrophysics and the evolution of the mathematical language.

Some new concepts were introduced such as the Nash flow generalization of the Ricci flow, the use of a general covariant affine connection totally consistent with

M. D. Maia, E. M. Monte, *Geometry at the Quantum Scale*,
https://doi.org/10.1007/978-3-031-61197-1

Einstein's principle of general covariance. We have seen that the concept of general covariance is not an option, but rather it is an inevitable consequence of the Einstein-Hilbert action, the derivation of Einstein's gravitation directly from a massless spin-2 quantum field, and the acknowledgment of the hyperbolic characteristics of Einstein's equations that warrants the legitimacy of gravitational waves. However, the most difficult problem that we have found was to convince our colleagues the assumption made by Planck and Einstein at around 1910 that the Newtonian Gravitational constant is not universal. Indeed that assumption remains the single factor that prevented the development of a unique theory of quantum gravity for more than a century. Once this difficulty is understood and removed, Einstein's relativistic philosophy can be understood as a renormalizable theory of gravitation without hierarchies.

The existence of quantum gravity based on Einstein's Relativity implies in the existence of a quantum four-dimensional space-time, similar to the classical space-time, with the difference that the observers cannot live in it. Then, and similarly to what happened with quantum field theory, we the observers remain confined in the classical space-time may adopt the diffeomorphism invariant affine connection as an observational tool to quantum gravity.

Finally we have found two interesting coincidences: First, the total dimensionality of the overall quantum space coincides with the same dimensionality of M-theory. Secondly, the Einstein-Hilbert action principle for the Standard Model in 11 dimensions can also reduce to four dimensions, suggesting a generalization of Donaldson's four-manifold theory to 11 dimensions.

References

1. James Clerk Maxwell, *A Treatise on Electricity and Magnetism*, First published (1873), Dover edition (1954).
2. Immanuel Kant, *Critique of Pure Reason*, (first published in 1781), English translation by J.M.D. Meiklejohn (1900), Chapter 2, section 2, p75, Dover (2003).
3. J. D. Jackson, *Electrodynamics* John Wiley and Sons, sixth print (1967).
4. G. Stephenson and C. W. Kilmister, *Special Relativity for Physicists* Dover Publications, 2019. ISBN 0486836606, 9780486836607
5. H. Minkowski *Space and Time* (A translation of an address delivered at the 80th Assembly of German Natural Scientists and Physicians, Cologne (1908), in A. Einstein, H. Minkowski and H. Weyl *The Principle of Relativity*, Notes by A. Sommerfeld. Dover Publications, New York (1923).
6. R. Penrose, *An Analysis of the Structure of Space-Time*. Adams Prize Essay, Cambridge University Press U.K. (1966).
7. Lie, Sophus (1888), *Theorie der Transformationsgruppen* I, Leipzig. Also Joel Merker, arXiv:1003.3202. ISBN 978-3-662-46210-2.
8. E. Wigner, *On the Consequences of the Symmetry of the Nuclear Hamiltonian on the Spectroscopy of Nuclei*. Phys. Rev. **51**, 106 (1937).
9. N. Hammermesh, *Group Theory and Its Application to Physical Problems*, Dover (1962).
10. V. S. Varadarajan, *Lie groups, Lie algebras, and their representations*, Prentice Hall (1975).
11. Bjorken and Drell *Relativistic Quantum fields*. McGrawll-Hill (1965)
12. J. Leite Lopes, *Classical Symmetries*. Latin American School of Physic, Caracas, Venezuela (1966). Also available from the Library of the Centro Brasileiro de Pesquisas Fisicas (CBPF) Rio de Janeiro, Brazil.
13. S. Weinberg, Phys. Rev. Lett. **19**, 1264 (1967).
14. A. Salam, Proc. 8th Nobel Simposium (Almquist Forlag) (1968).
15. M. Gell-Mann and Y. Ne'eman, *The Eightfold Way* Benjamin (1964)
16. Y. Ne'emann, *Derivation of strong interactions from gauge invariance*. Nuclear Physics *26*, 222 (1961).
17. F.J. Dyson, *Symmetry Groups in Nuclear and Particle Physics*, Benjamin (1966).
18. A. Pickering, *Constructing Quarks*. University of Chicago (1984)
19. C. N. Yang and R. Mills, *Conservation of Isotopic Spin and Isotopic Gauge Invariance*. Phys. Rev. *96*, 191 (1954). https://doi.org/10.1103/PhysRev.96.191.
20. Dreschsler & Mayer, *Fiber Bundle Techniques in Gauge Theories*. Lecture Note in Physics. Springer (1977).

M. D. Maia, E. M. Monte, *Geometry at the Quantum Scale*,
https://doi.org/10.1007/978-3-031-61197-1

21. S. K. Donaldson *An application of gauge theory to Four Dimensional Topology*. J. Differential Geometry **18**, 279–318 (1983)
22. S. K. Donaldson, *Smooth 4-manifolds with definite Intersection forms*. Contemporary Mathematics, AMS, **35**, 201 (1984)
23. V. M. Mostepanenko and G. L. Klimchitskaya. *The State of the Art in Constraining Axion-to-Nucleon Coupling and Non-Newtonian Gravity from Laboratory Experiments. Universe.* **6**, 147 (2020).
24. L.P. Eisenhart, *Riemannian Geometry* Princeton U. P. (1925).
25. Barrett O'Neil *Elementary Differential Geometry*. Academic Press (1996)
26. William M. Boothby. *Introductions to Differentiable Manifolds and Riemannian Geometry*, Academic Press (1975).
27. G. Ricci and T. Levi-Civita, *Methodes de Calcul Differentiel Absolu et Leurs Applications*. Matematische Annalem, **54**, 125–201 (1901).
28. Hans Stephani, Dietrich Kramer, Malcolm MacCallum, Cornelius Hoenselaers, Eduard Herlt. *Exact Solutions of Einstein's Field Equations*. Cambridge University press (2003).
29. F. A. E. Pirani, *On the Physical Significance of the Riemann Tensor*, Acta Physica Polonica **15**, 389 (1956).
30. M. Fierz and W.Pauli *On Relativistic Wave Equations for Particles of Arbitrary Spin in an Electromagnetic Field*. Proc. Roy. Soc. Lond. A **173**, 211 (1939).
31. A.Z. Petrov (1954). *Classification of spaces defined by gravitational fields*. Uch. Zapiski Kazan. Gos. Univ. 114 (8), 55.
32. A.Z. Petrov (1969). Einstein Spaces. Oxford: Pergamon. ISBN 0080123155.
33. Ezra T. Newman and Roger Penrose. *An Approach to Gravitational Radiation by a Method of Spin Coefficients*. Journal of Mathematical Physics **3**, 566 (1962). See also Ezra T Newman, and Roger Penrose. Errata: *An Approach to Gravitational Radiation by a Method of Spin Coefficients*. Journal of Mathematical Physics **4**, 998 (1963)
34. A. Friedman (1999). *On the Curvature of Space". General Relativity and Gravitation* **31**, 1991 (1922). Translation from the original: A. Friedman, *Uber die Krummung des Raumes*. Z. Phys. 10(1): 377.
35. S. Weinberg Gravitation and Cosmology: Principles and Applications of the General Theory of Relativity. J. Willey (1972).
36. M. Milgrom, *A modification of the Newtonian dynamics as a possible alternative to the hidden mass hypothesis*. Astrophysical Journal **270**, 365 (1983).
37. Fritz Zwicky, Helvetica Physica Acta 6: 110 (1933).
38. Vera Rubin, & W. K. Ford, Jr. Astrophysical Journal **159**, 379 (1970). https://doi.org/10.1086/150317
39. E. Inonu and E. F. Wigner, *On the Contraction of Groups and Their Representations*. Proc. Natl. Acad. Sci. **39**, 510 (1953).
40. Ya. B. Zel'dovich and I. D. Novikov. *The Structure and Evolution of the Universe*. Chicago U. P. (1983).
41. R. A. Hulse and J. H. *A High-Sensitivity Pulsar Survey*. Astrophys. J. **191**, L59 (1974)
42. The LIGO Scientific Collaboration and Virgo Collaboration. *Observation of Gravitational Waves from a Binary Black Hole Merger*. Physical Review Letters 116, 061102 (2016).
43. A. Trautmann, F.A. E. Pirani and H Bondi *Lectures on General Relativity*. Brandeis Summer Institute in Theoretical Physics, Prentice Hall (1965). London.
44. P. Szekeres, Nuovo Cimento. **43**, 1062 (1966).
45. S. Carlip and others *Quantum Gravity: A brief History of Ideas and some prospects*, arXiv:1507.08194 (2015).
46. Martin Bojowald *Quantum Cosmology. A Fundamental Description of the Universe*. Lecture Notes in Physics **835**. Springer. ISBN 978-1-4419-8276-6, arXiv:1501.04899v1 [gr-qc].
47. R. Penrose, *Angular Momentum: an Approach to Combinatorial Spacetime*, in *Quantum Theory and Beyond*. Cambridge University Press, T. Bastin (ed.). Academic Press, 221–244 (1971).

48. C. J. Isham, R. Penrose and D. W. Sciama, *Quantum Gravity*. Clarendon Press, Oxford (1975).
49. Abbay Ashtekar, *New Variables in Classical and Quantum gravity*. Phys. Rev. Lett. 57, 2244 (1986).
50. Lee Smolin, *An Invitation to loop quantum gravity*, in Quantum theory and symmetries. Cincinnati, 655–682 (2003). hep-th/0408048.
51. Lee Smolin, *The Trouble with Physics*. Houghton-Mifflin (2006).
52. G. 't Hooft and M. Veltman, *One-loop divergences in the theory of gravitation*. Ann. Inst. Henri Poincaré **20**, 69 (1974).
53. A. Sommerfeld, Partial Differential Equations on Theoretical Physics, vol VI. Lecture notes in physics, Academic Press (1964)
54. Y. Choquet Bruhat and Tommaso Ruggeri. *Hyperbolic 3 + 1 System of Einstein's Equations*. Comm. Math. Phys. **89**, 269 (1983).
55. Lang, S. (1995). Differential and Riemannian manifolds. *The Theorem of Frobenius*. Springer-Verlag. ISBN 978-0-387-94338-1.
56. Perter D. Lax, *Functional Analysis*. Wiley-Interscience (2002). ISBN:0471556041.
57. J. York Jr. (2004,5). *The Initial Value Problem Using Metric and Extrinsic Curvature*, Proceedings of the 10th Marcel Grossmann Meeting on Recent Developments in Theoretical and Experimental General Relativity, Gravitation and Relativistic Field Theories, 3–16, arXiv:0405005 [gr-qc].
58. K. Kuchar, *Time and Interpretations of Quantum Gravity*. Proceedings of the 4th Canadian Conference on General Relativity and Relativistic Astrophysics. G. Kunstatter, D. Vincent, and J. Williams Eds. World Scientific, Singapore (1992)
59. P.A.M. Dirac, *Fixation of Coordinates in the Hamiltonian Theory of Gravitation*. Phys. Rev. **114**, 924 (1959).
60. J.N. Goldberg, *Triad Approach to the Hamiltonian of General Relativity*. Phys. Rev. **D37**, 2116–2120 (1988).
61. J. Ambjorn, J. Jurkiewicz, R. Loll, *Quantum Gravity, or The Art of Building Spacetime*, Spin-06/16, ITP-UU-06/19, arXiv:hep-th/0604212v1 (2006).
62. T. P. Shestakova, *Hamiltonian Formulation of General Relativity 50 Years after the Dirac Celebrated Paper: Do unsolved problems still exists?*, arXiv:0911.5252v2 [gr-qc]
63. S. Mandelstam. *Analytic Properties of Transition Amplitudes in Perturbative Theory*. Physical Reviews **115**, 1741 (1959).
64. H. Weyl, Bull. Am. Math. Sot. 44, 598 (1938)
65. E. Cartan, *Lessons sur les espace de Riemann*. Gauthier-Villars, Paris (1946).
66. P. Griffiths, Duke Math. J. 41, 775 (1974).
67. G. R. Jensen, *Higher-order Contact of Submanifolds of Homogeneous Spaces*. Lecture Notes in Mathematics, vol. 610 (Springer-Verlag, Berlin, 1977).
68. R. Arnowitt, S. Deser and C. W. Misner *The Dynamics of General Relativity*, in *Gravitation: an introduction to current research*, L. Witten (ed.), pp. 227–264. Wiley, New York (1962). Also in arXiv:gr-qc/0405109v1
69. M. D. Maia, J. Math. Phys. 22, 538 (1981)
70. M. D. Maia and W. L. Roque. *High order space-time contacts and equivalence*. J. Math. Phys., Vol. 32, No. 10, October (1991)
71. H. Bethe, Phys. Rev. vol. 72 339 (1947)
72. K. G. Wilson and J. Kogut The Renormalization Group and the ε Expansion. Physics Reports **12c** (1974)
73. K. G. Wilson, *Problems in physics with many scales of length*, Scientific American, August 1979 "Archived copy" (PDF). Archived from the original (PDF) on 2012-07-10. Retrieved 2013-06-18.
74. K. G. Wilson. *The renormalization group and critical phenomena*. Reviews of Modern Physics 55(3): 583 (1983)
75. Kerson Huang, *Fundamental Forces of Nature (a story of Gauge Fields)*, World Scientific Publishing Co. (2007)

76. Kerson Huang, *A critical History of Renormalization*. Int. J. Mod. Phys. A, 28 (2013) 1330050, arXiv:1309.5707.
77. Michael E. Peskin and Daniel V. Schroeder, *Introduction to quantum field theory*. Perseus Books, The Advanced Book Program Reading, Massachusetts, EUA (1951).
78. G. 't Hooft and M. Veltman, Universitext *Regularization and Renormalization of Gauge Fields*. Nucl. Phys. **B44**, 189 (1972).
79. V. N. Popov, *Functional Integral in Quantum Field Theory*. Lecture Notes Theory Division Th2424-CERN, December (1977)
80. R. Fueter, Commentarii Mathematici Helvetici, **10**, 306 (1927)
81. Sheldon Glashow, *Towards a Unified Theory: Threads in a Tapestry (Nobel Lecture)*; Review of Modern Physics, Vol. 52, Issue 3: 539 (1980)
82. Goroff, M. and Sagnotti, A. *Quantum gravity at two loops*. Physics Letters **B160**, 81 (1985). https://doi.org/10.1016/0370-2693(85)91470-4
83. M. Planck, *Theory of Heat Radiation*. Second edition (1907). Blakiston's Son & Co, Philadelphia.
84. E. M. Lifschitz, *On the Theory of Second-Order Phase Transitions I and II*. Zh. Eksp. Teor. Fiz. **11**, 255, 269 (1941).
85. D. Wotton, *The Invention of Science*, Chapter 5: The Mathematization of the World. Penguin, Random House (2015)
86. Markus Fierz. *Uber die relativistische Theorie kraftefreier Teilchen mit beliebigem Spin*. Helvetica Physica Acta. **12**(1), 3 (1939). Bibcode:1939AcHPh..12....3F. https://doi.org/10.5169/seals-110930
87. Wolfgang Pauli *The Connection Between Spin and Statistics*. Physical Review **58**, 716 (1940).
88. David G. Boulware, Stanley Deser. *Can gravitation have a finite range?*. Phys. Rev. D. 6, 3368 (1972).
89. M. D. Maia *Cosmology of Spin-2 Fields*, in Selected Papers from the Ninth Alexander Friedmann Seminar on Gravitation and Cosmology and the third Symposium on the Casimir Effect. International Journal of Modern Physics A, Saint Petersburg, Russia **31**, 1641010-1–1641010-10, World Scientific (2016).
90. Matvei Petrovich Bronstein, *Quantum theory of weak gravitational fields*. Phys. Zeitschr. der Sowjetunion **9**, 140 157 (1936). M.P. Bronstein, arXiv:1110.5941 (1936).
91. S. Deser and A. Starobinsky *Introduction to Bron stein's Quantum theory of weak gravitational fields. A scientific introduction and short biography*. Gen. Rel. Grav. **44**, 263 (2012), arXiv1110.5941
92. S. Deser, *A short Pre-history of Quantum Gravity*. Report number: brx-th-6697; caltech-th-2022-001, arXiv2201.05736[gr-qc] (2022).
93. *Feynman's Lectures on Gravitation* Foreword by John Preskill and Kipp S. Thorne in (1995) Feynman, R. Fernando Morinigo, William Wagner, Brian Hatfield, David Pines. Addison Wesley Publishing Company. Menlo Park, CA, USA.
94. Suraj N. Gupta, *Quantization of Einstein's Gravitational Field: General Treatment*. Proceedings of the Physical Society. Series A. 65(8): 608 (1952). Bibcode:1952PPSA...65..608G. https://doi.org/10.1088/0370-1298/65/8/304.
95. Suraj N. Gupta, *Gravitation and Electromagnetism*. Phys. Rev. 96, 1683 (1954). https://doi.org/10.1103/PhysRev.96.1683
96. R. H. Kraichnan. *Special Relativistic Derivation of Generally Covariant Gravitation Theory. Physical Reviews* **98**, 1118 (1955).
97. Anthony Bayes *An Essay Towards Solving a Problem in the Doctrine of Chances*, Philosophical Transactions, of the Royal Society of London **53**, 370 (1764).
98. P T. Padmanabhan *From Graviton to Gravity: Myths and Reality*, arXiv:gr-qc/0409089v1 23 Sep. 2004
99. Valentin I. Zakharov, *Linearized gravitation theory and the graviton mass*. JETP Lett. 12: 312 (1970).
100. Hendrik van Dam; Martinus J. G. Veltman (1970). *Massive and massless Yang-Mills and gravitational fields*. Nucl. Phys. B. 22(2): 397

101. A.I. Vainshtein, *To the problem of nonvanishing gravitation mass*. Phys. Lett. B. 39(3): 393 (1972).
102. Claudia de Rham, *Massive Gravity*, arXiv:1401.4173v2 [hep-th] 14 Mar 2014.
103. *Review of Particle Physics*, Physical Review D **98**, 3 (2018).
104. M. Creutz, L. Jacobs and C. Rebbi *Monte Carlo computations in Lattice Gauge Theories*, Physics Reports **95**, 4 (1983).
105. G. Ricci and T. Levi-Civita, *Methodes de calcul Differentiel Absolu et Leurs Applications*. Matematische Annalem, vol 54, 125–201 (1901).
106. Christoffel, E.B., *Ueber die Transformation der homogenen Differentialausdrucke zweiten Grades*. Journal fur die Reine und Angewandte Mathematik, B. **70**, 46 (1869). https://doi.org/10.1515/crll.1869.70.46
107. E. Cartan, *Sur les varietes a Conexion affine et la theorie de la Relativite Generalisee*. Ann. Ecole Normale Sup **40** (1923).
108. B. Riemann, *Ueber die Hypothesen welche der Geometrie zu Grunde liegen*, Gesammelte Matematische Werke, 272–287, Leipzig (1892).
109. a. Einstein, *Die Grundlage der algemeinen Relativitatstheorie*, Annalen der Physik (1916).
110. S. Coleman, J. Mandula, *All possible symmetries of the S matrix*. Phys. Rev. 159, 1251 (1967)
111. S. P. Rosen. *Finite Transformations in Various Representations of SU(3)*. Journal of Mathematical Physics **12**(4), 673 (1971).
112. T L Curtright, C. K. Zacho. *Elementary results for the fundamental representation of SU(3)*. Reports on Mathematical Physics **76**(3), 401 (2015), arXiv:1508.00868.
113. E. Witten, *Search for a realistic Kaluza-Klein theory*. Nuclear Physics **B 180**, 412 (1981).
114. L. O'Raifeartaigh, Phys. Rev. Lett. 14, 475 (1965)
115. L. O'Raifeartaigh, Phys. Rev. **139B**, 1052 (1965).
116. Eugene Wigner, *On unitary representations of the inhomogeneous Lorentz group*. Annals Math. **40**, 149 (1939).
117. Hendrik Brugt Gerhard Casimir. *Rotation of a rigid body in quantum mechanics*. Groningen, den Haag, Batavia (1931).
118. O. W. Greenberg, Phys. Rev. **235**, 1443 (1964)
119. L. Michel, Phys. Rev. **137**, B 405 (1965).
120. M. Flato, O. Sternheimer, Phys. Rev. Lett. **15**, 934 (1965).
121. I. Segal J. Functional Analysis, 1, 1 (1967).
122. R. Haag, J.T. Lopuszanski, M. Sohnius, Nuclear Physics B 86, (1975).
123. P. van Nieuwenhuizen, *Supergravity*. Physics reports **68**, 4 (1981).
124. P. Fayet and S Ferrara, *Supersymmetry*, Physics Reports 32, 249–334 (1977).
125. John H. Schwarz *Superstring Theory*. Physics Reports **89** (1982).
126. Edward Witten, *Strong Coupling Expansion Of Calabi-Yau Compactification*, Nucl. Phys. B 471:135–158 (1996) (arXiv:hep-th/9602070).
127. David J. Gross, Jeffrey A. Harvey, Emil Martinec, Ryan Rohm, *Heterotic String*. Physical Review Letters. **54**, 502 (1985).
128. Petr Horava, Edward Witten, *Heterotic and type I string dynamics from eleven dimensions*, Nucl. Phys. B 460 (1996), (arXiv:hep-th/9510209).
129. Petr Horava, Edward Witten, *Eleven dimensional supergravity on a manifold with boundary*, Nucl. Phys. B475 (1996) 94 (arXiv:hep-th/9603142).
130. Juan M. Maldacena, *The Large N Limit of Superconformal Field Theories and Supergravity*. Adv. Theor. Math. Phys. **2**, 231 (1998).
131. Joseph Polchinski. *String Theory: Superstring Theory and Beyond*. Cambridge University Press, pp. 55 (1998). ISBN 9780521633048.
132. E. Cunningham, *Transformation of the electromagnetic equations* Proc. London Math. Soc. **6** 77 (1909)
133. H. Bateman *The transformations of coordinates which can be used to transform one physical problem into another* Proc. Roy. Soc. Math. **8**, 223 (1910)
134. P. Higgs, *Broken Symmetries and the Masses of Gauge Bosons*. Physical Review Letters **13**(16), 508 (1964).

135. N. Hammermesh, *Group Theory and its Application to Physical Problems*, Dover (1962).
136. M. Gell-Mann and Y. Ne'emann, *The Eightfold Way: A theory of strong interaction symmetry*. Synchrotron Laboratory Report CTSL-20 (California Institute of Technology), (1961). The Eightfold Way, Benjamin (1964)
137. Anthony W. Knap *Lie Groups Beyond an Introduction*. Springer Series: Progress in mathematics (Boston, Mass.), vol. 140.
138. G. Fano, *Sui postulati fondamentali della geometria proiettiva*, Giornale di Matematiche, Verona, Italy **30**, 106 (1891). Biblioteca Digitale Italiana di Matematica http:
139. Baez, John C. and Huerta, John. *Division Algebras and Supersymmetry I*. R. Doran, G. Friedman, J. Rosenberg, (eds.). *Superstrings, Geometry, Topology, and C*-algebras* (2010) American Mathematical Society, arXiv:0909.0551.
140. M. Gunaydin & F. Gursey, *Quark Structure and Octonions*, J. Math. Phys. **14**, 1651 (1973).
141. M. Gunaydin, *Octonionic Hilbert space, the Poincaré group and SU(3)*. J. Math. Phys. **17**, 1875 (1976).
142. Theodor Kaluza, Proceedings of the Prussian Academy of Sciences. archive.org. pp. 966 (1918).
143. Theodor Kaluza, *Zum Unitatsproblem in der Physik* Sitzungsber Preuss. Akad. Wiss. Berlin. (Math. Phys.) 966 (1921).
144. O. Klein, *Quantentheorie und funfdimensionale Relativitastheorie*. Zeitschrift fur Physik. **37**, 12 895 (1926).
145. Albert Einstein, P. Bergmann *On a Generalization of Kaluza's Theory of Electricity*. Annals Math. **39** 683 (1938).
146. C. DeWitt and B. DeWitt, *Dynamical Theory of Groups and Fields*, in *Relativity, Groups & Topology*: Proceedings of the Les Houches Summer School, Gordon and Breach (1963).
147. J. Nash, *The Imbedding Problem for Riemannian Manifolds*. Annals of Mathematics **63**, 20 (1956).
148. R. Greene, *Isometric embeddings of Riemannian and pseudo-Riemannian manifolds*. Memoirs of the American Mathematical Society, number 97 (1970). ISBNs: 978–8218-1297-6 (print); 978-1-4704-0047-7 (online). https://doi.org/10.1090/memo/0097
149. Femilab *The Muon g2 Experiment*. The Fermi National Accelerator Laboratory (Fermilab), High-Energy Particle Physics Laboratory, United States Department of Energy (2017). Batavia, Illinois, USA.
150. Edmundo M. Monte and M. D. Maia, *The Twisting Connection of Space-time*. Journal of mathematical Physics **37**, 4 (1996).
151. M. D. Maia, E. M. Monte, J. M. F. Maia and J. S. Alcaniz *On the Geometry of Dark Energy*. Classical and Quantum Gravity, **22**, 1623 9 (2005). https://doi.org/10.1088/0264-9381/22/9/010
152. James Lindsey, Reza Tavakol and C. Romero *Campbell's Embedding Theorem*. October 1997 Modern Physics Letters A 12(31), 2319–2323 (1997).
153. F. Dahia and C. Romero *The embedding of the space–time in five dimensions:* An extension of the Campbell–Magaard theorem. J. Math. Phys. **43**, 5804 (2002).
154. A. Voss, *Zur theorie de transformation QuadratischerDifferentialausdruke and der Krummung Hóherer Mannifaltigkeiten.* Matematische Annalen vol 16, 129 (1880).
155. Henri Poincaé. *A closed, smooth, simply connected 3-manifold is diffeomorphic to* S^3 in *Cinquieme comment a l'analysis situs*. Gauthier Villars, Great Classics, pages 541. Editions Jacques Gabay, Sceaux (1996).
156. J. W. Morgan and G. Tien *Ricci flow and the Poincare conjecture* AXiv:math/0607607 21 Mar (2007)
157. R. Hamilton, J. Diff. Geom. **17**, 255 (1982), G. Perelman, arXiv:math/8211159
158. G. Perelman, *The entropy formula for the Ricci flow and its geometric applications*. e-print: arXiv:math.DG/0211159 (2002)
159. G. Perelman, *Ricci flow with surgery on three-manifolds*. e-print: arXiv:math.DG/0303109 (2003). Also G. Perelman, *Finite extinction time for the solutions to the Ricci flow on certain three-manifolds*. e-print: arXiv:math.DG/0307245 (2003).

160. Nasar, Sylvia; Gruber, David. *Manifold Destiny*: A legendary problem and the battle over who solved it. The New Yorker, 21 August 2006. N.Y. USA.
161. Yano, K. (1957). *The Theory of Lie Derivatives and its Applications*. North-Holland. p. 8. ISBN 978-0-7204-2104-0.
162. M. Crampin and F.A. E. Pirani, *Applicable Differential Geometry*, Cambridge UP (1986).
163. Edmundo M. Monte and M. D. Maia *A Fundamental Theorem of Space-times*. 330arXiv.org/abs/gr-qc/9512028v1 (1995).
164. Guralnik, G.; Hagen, C. R.; Kibble, T. W. B. "Global Conservation Laws and Massless Particles". Physical Review Letters. 13(20): 585 (1964). Bibcode:1964PhRvL..13..585G. https://doi.org/10.1103/PhysRevLett.13.585.
165. Brout, R.; Englert, F. (1998). "Spontaneous Symmetry Breaking in Gauge Theories: A Historical Survey". arXiv:hep-th/9802142.
166. Leon Lederman & Dick Teresi, *The God Particle. If The universe is the answer, what is the question?* Houghton Mifflin Harcourt Publishing Company (1993)
167. Eugene Wigner, *Group Theory and its Application to the Quantum Mechanics of Atomic Physics*. Academic Press, New York (1959)
168. J.R. Oppenheimer, H. Snyder *On Continued Gravitational Contraction*. Phys. Rev. 56 455–459 (1939)
169. Finkelstein, D. *Past-Future Asymmetry of the Gravitational Field of a Point Particle*. Physical Review **110**(4): 965 (1958).
170. M. Kruskal, *Maximal Extension of Schwarzschild Metric*. Physical Review **119**(5), 1743 (1960). Bibcode:1960PhRv..119.1743K. https://doi.org/10.1103/PhysRev.119.1743.
171. S. W. Hawking. *Gravitationally collapsed objects of very low mass*. Monthly Notices of the Royal Astronomical Society. 152: 75 (1971). Bibcode:1971MNRAS.152...75H. https://doi.org/10.1093/mnras/152.1.75
172. S. W. Hawking. B *lack hole explosions?*. Nature. 248 (5443): 30–31 (1974). Bibcode:1974Natur.248...30H. https://doi.org/10.1038/248030a0. ISSN 1476–4687. S2CID 4290107
173. S. W. Hawking, *Particle Creation by Black Holes*. Communications in Mathematical Physics. 43 (3): 199 (1975). bibcode:1975CMaPh..43..199H. https://doi.org/10.1007/BF02345020. S2CID 55539246.
174. A. Bekenstein. *Black holes and the second law*. Lettere al Nuovo Cimento. 4(15): 99 (1972). https://doi.org/10.1007/BF02757029.
175. J. D Bekenstein, *Can quantum gravity be exposed in the laboratory?: A tabletop experiment to reveal the quantum foam*. Found. Phys. 44, 452–462 (2014). arXiv:1211.3816
176. E. Winstanley, *Black Holes, Tev-Scale Gravity and the LHC*, Time and Matter Conference, Venice, Italy (2013), arXiv:1306.5409v1 [hep-th]
177. Javad T. Firouzjaee and George E. R. Ellis, *Particle creation Rate for Dynamical Black Holes* (2016), arXiv:1511:04316v3 [gr-qc]
178. G. C. Wick. *Properties of Bethe-Salpeter Wave Functions*. Physical Review. **96**(4), 1124 (1954). Bibcode:1954PhRv...96.1124W. https://doi.org/10.1103/PhysRev.96.1124.
179. Anthony Zee, *Quantum Field Theory in a Nutshell* (2nd ed.). Princeton University Press, p. 289 (2010). ISBN 978-1-4008-3532-4.
180. M. Janet, Ann. Soc. Polon. Math. **5**, 38 (1926).
181. E. Cartan, An. Soc. Polon. Math. **6**, 1 (1927).
182. S. W. Hawking *Black holes and the information paradox*. Proc. of the 17th International Conference on General Relativity and Gravitation. Petros Florides, Brien Nolan and Adrian Ottwell (eds). World Scientific, Singapore, 56–62 (2004)
183. Nima Arkani-Hamed, Yu-tin Huang, Donal O'Connell, *Kerr black holes as elementary particles*. JHEP 01 046 (2020), ArXix: 1906.10100
184. Martinus Veltman *Diagrammatica*, Cambridge University Press Print publication (994). Also available online 2011. Online ISBN:9780511564079. https://doi.org/10.1017/CBO978051156407

185. Thomas Gutierrez, *the deconstructed standard model equation* Symmetry Magazine (2016). https://www.symmetrymagazine.org/article/the-deconstructed standard model equation
186. W. Drechsler, and M. E. Mayer, *Fiber Bundle Techniques in Gauge Theory* Lecture Notes in Physics **67** (1977). Springer-Verlag, Berlin.
187. M. Scher, *Electroweak Higgs Potentials and Vacuum Stability*, Physics Reports **179**, 273 (1989)
188. D. B. Lichtemberg, *Unitary Symmetry and Elementary Particles*. Academic Press (1978)
189. M. D. Maia, J. S. Alcaniz, Edmundo M. Monte, A. J. S. Capistrano, *The deformable Universe*. General Relativity and Gravitation, (**43**), 2685–2700 (2011). https://doi.org/10.1007/link.springer.com/article/10.1007/s10714-011-1192-8
190. J. F. Donoghue Phys.Rev.D50:3874-3888,1994. General Relativity and Quantum Cosmology (gr-qc); High Energy Physics—Phenomenology (HEP-PH); High Energy Physics—Theory (hep-th), arXiv:gr-qc/9405057–ce.

Index

M. D. Maia, E. M. Monte, *Geometry at the Quantum Scale*,
https://doi.org/10.1007/978-3-031-61197-1

MIX
Papier aus verantwortungsvollen Quellen
Paper from responsible sources
FSC® C105338

If you have any concerns about our products,
you can contact us on
ProductSafety@springernature.com

In case Publisher is established outside the EU,
the EU authorized representative is:
Springer Nature Customer Service Center GmbH
Europaplatz 3, 69115 Heidelberg, Germany

Printed by Libri Plureos GmbH
in Hamburg, Germany